Nanostructured Materials and Systems

Nanostructured Materials and Systems

Ceramic Transactions, Volume 214

A Collection of Papers Presented at the 8th Pacific Rim Conference on Ceramic and Glass Technology May 31–June 5, 2009 Vancouver, British Columbia

Edited by
Sanjay Mathur
Hao Shen

Volume Editor
Mrityunjay Singh

The American Ceramic Society

A John Wiley & Sons, Inc., Publication

Published by John Wiley & Sons, Inc., Hoboken, New Jersey.
Published simultaneously in Canada.

No part of this publication may be reproduced, stored in a retrieval system, or transmitted in any form
or by any means, electronic, mechanical, photocopying, recording, scanning, or otherwise, except as
permitted under Section 107 or 108 of the 1976 United States Copyright Act, without either the prior
written permission of the Publisher, or authorization through payment of the appropriate per-copy fee to
the Copyright Clearance Center, Inc., 222 Rosewood Drive, Danvers, MA 01923, (978) 750-8400, fax
(978) 750-4470, or on the web at www.copyright.com. Requests to the Publisher for permission should
be addressed to the Permissions Department, John Wiley & Sons, Inc., 111 River Street, Hoboken, NJ
07030, (201) 748-6011, fax (201) 748-6008, or online at http://www.wiley.com/go/permission.

Limit of Liability/Disclaimer of Warranty: While the publisher and author have used their best efforts in
preparing this book, they make no representations or warranties with respect to the accuracy or
completeness of the contents of this book and specifically disclaim any implied warranties of
merchantability or fitness for a particular purpose. No warranty may be created or extended by sales
representatives or written sales materials. The advice and strategies contained herein may not be
suitable for your situation. You should consult with a professional where appropriate. Neither the
publisher nor author shall be liable for any loss of profit or any other commercial damages, including
but not limited to special, incidental, consequential, or other damages.

For general information on our other products and services or for technical support, please contact our
Customer Care Department within the United States at (800) 762-2974, outside the United States at
(317) 572-3993 or fax (317) 572-4002.

Wiley also publishes its books in a variety of electronic formats. Some content that appears in print may
not be available in electronic format. For information about Wiley products, visit our web site at
www.wiley.com.

Library of Congress Cataloging-in-Publication Data is available.

ISBN 978-0-470-88128-6

Printed in the United States of America.

10 9 8 7 6 5 4 3 2 1

Contents

Preface

The Symposium on *Nanostructured Materials and Systems* was held during the 8th Pacific Rim Conference on Ceramic and Glass Technology (PACRIM 8) from May 31–June 5, 2009 in Vancouver, Canada. This symposium aimed to review the progress in the state-of-the-art of nanoscience and nanotechnology including synthesis, processing, modeling, applications and assessment of toxicological potential of nanomatter. More than 55 contributions (invited talks, oral presentations, and posters), were presented by participants, from all over the world, representing universities, research institutions, and industry which made this symposium an attractive forum for interdisciplinary presentations and discussions and to elaborate their functional diversity.

This issue contains 16 peer-reviewed papers (invited and contributed) incorporating the latest developments related to synthesis, processing and manufacturing technologies of nanoscaled materials and systems including one-dimensional nanostructures, nanoparticle-based composites, electrospinning of nanofibers, functional thin films, ceramic membranes, bioactive materials and self-assembled functional nanostructures and nanodevices. These papers discuss several important aspects related to fabrication and engineering issues necessary for understanding and further development of processing and manufacturing of nanostructured materials and systems.

We would like to thank all members of the Organizing Committee for their tireless work in the planning and execution of the symposium in a meticulous manner. The editors wish to extend their gratitude and appreciation to all the authors for their contributions, to all the participants and session chairs for their time and efforts, and to all the reviewers for their valuable comments and suggestions. Financial support from Plasma Electronic GmbH, Germany and the Engineering Ceramics Division of The American Ceramic Society is gratefully acknowledged. The invaluable assistance of the staff of the meeting and publication departments of The American Ceramic Society is gratefully acknowledged.

We hope that the collection of papers presented here will serve as a useful reference for the researchers and technologists working in the field of interested in science and technology of nanostructured materials and devices.

SANJAY MATHUR
HAO SHEN
University of Cologne
Cologne, Germany

Introduction

The 8th Pacific Rim Conference on Ceramic and Glass Technology (PACRIM 8), was the eighth in a series of international conferences that provided a forum for presentations and information exchange on the latest emerging ceramic and glass technologies. The conference series began in 1993 and has been organized in USA, Korea, Japan, China, and Canada. PACRIM 8 was held in Vancouver, British Columbia, Canada, May 31–June 5, 2009 and was organized and sponsored by The American Ceramic Society. Over the years, PACRIM conferences have established a strong reputation for the state-of-the-art presentations and information exchange on the latest emerging ceramic and glass technologies. They have facilitated global dialogue and discussion with leading world experts.

The technical program of PACRIM 8 covered wide ranging topics and identified global challenges and opportunities for various ceramic technologies. The goal of the program was also to generate important discussion on where the particular field is heading on a global scale. It provided a forum for knowledge sharing and to make new contacts with peers from different continents.

The program also consisted of meetings of the International Commission on Glass (ICG), and the Glass and Optical Materials and Basic Science divisions of The American Ceramic Society. In addition, the International Fulrath Symposium on the role of new ceramic technologies for sustainable society was also held. The technical program consisted of more than 900 presentations from 41 different countries. A selected group of peer reviewed papers have been compiled into seven volumes of The American Ceramic Society's Ceramic Transactions series (Volumes 212-218) as outlined below:

- **Innovative Processing and Manufacturing of Advanced Ceramics and Composites, Ceramic Transactions, Vol. 212,** Zuhair Munir, Tatsuki Ohji, and Yuji Hotta, Editors; Mrityunjay Singh, Volume Editor
  Topics in this volume include Synthesis and Processing by the Spark Plasma

Method; Novel, Green, and Strategic Processing; and Advanced Powder Processing

- **Advances in Polymer Derived Ceramics and Composites, Ceramic Transactions, Vol. 213,** Paolo Colombo and Rishi Raj, Editors; Mrityunjay Singh, Volume Editor
  This volume includes papers on polymer derived fibers, composites, functionally graded materials, coatings, nanowires, porous components, membranes, and more.

- **Nanostructured Materials and Systems, Ceramic Transactions, Vol. 214,** Sanjay Mathur and Hao Shen, Editors; Mrityunjay Singh, Volume Editor
  Includes papers on the latest developments related to synthesis, processing and manufacturing technologies of nanoscale materials and systems including one-dimensional nanostructures, nanoparticle-based composites, electrospinning of nanofibers, functional thin films, ceramic membranes, bioactive materials and self-assembled functional nanostructures and nanodevices.

- **Design, Development, and Applications of Engineering Ceramics and Composite Systems, Ceramic Transactions, Vol. 215,** Dileep Singh, Dongming Zhu, and Yanchun Zhou; Mrityunjay Singh, Volume Editor
  Includes papers on design, processing and application of a wide variety of materials ranging from SiC SiAlON, ZrO_2, fiber reinforced composites; thermal/environmental barrier coatings; functionally gradient materials; and geopolymers.

- **Advances in Multifunctional Materials and Systems, Ceramic Transactions, Vol. 216,** Jun Akedo, Hitoshi Ohsato, and Takeshi Shimada, Editors; Mrityunjay Singh, Volume Editor
  Topics dealing with advanced electroceramics including multilayer capacitors; ferroelectric memory devices; ferrite circulators and isolators; varistors; piezoelectrics; and microwave dielectrics are included.

- **Ceramics for Environmental and Energy Systems, Ceramic Transactions, Vol. 217,** Aldo Boccaccini, James Marra, Fatih Dogan, Hua-Tay Lin, and Toshiya Watanabe, Editors; Mrityunjay Singh, Volume Editor
  This volume includes selected papers from four symposia: Glasses and Ceramics for Nuclear and Hazardous Waste Treatment; Solid Oxide Fuel Cells and Hydrogen Technology; Ceramics for Electric Energy Generation, Storage, and Distribution; and Photocatalytic Materials.

- **Advances in Bioceramics and Biotechnologies, Ceramic Transactions, Vol. 218;** Roger Narayan and Joanna McKittrick, Editors; Mrityunjay Singh, Volume Editor
  Includes selected papers from two cutting edge symposia: Nano-Biotechnology and Ceramics in Biomedical Applications and Advances in Biomineralized Ceramics, Bioceramics, and Bioinspiried Designs.

I would like to express my sincere thanks to Greg Geiger, Technical Content Manager of The American Ceramic Society for his hard work and tireless efforts in the publication of this series. I would also like to thank all the contributors, editors, and reviewers for their efforts.

MRITYUNJAY SINGH
Volume Editor and Chairman, PACRIM-8
Ohio Aerospace Institute
Cleveland, OH (USA)

HYDROGEN PERMEABLE MEMBRANES FROM PALLADIUM COATED ANODIC ALUMINA

Ian Brown, Jeremy Wu, Melanie Nelson, Mark Bowden, Tim Kemmitt
The MacDiarmid Institute for Advanced Materials and Nanotechnology, Industrial Research Ltd., P.O. Box 5040, Gracefield Road, Lower Hutt, New Zealand

ABSTRACT
        Nanostructured anodic alumina membranes have been fabricated from templates prepared by anodizing annealed high-purity aluminum foil. A range of acid electrolytes and anodizing voltage profiles have been used to control the pore diameter, pore separation and thickness in these alumina ceramic templates. Thin films of Pd metal up to 200nm thick have been deposited on the surface of the alumina templates and we have demonstrated that the membranes formed can be used to separate hydrogen from mixed gas streams with very high selectivity. These pure Pd membranes experience hydrogen embrittlement in operation below 350°C and we have employed silver as an alloying metal to help prevent this. An electroless deposition technique was used to obtain thin films of palladium and silver on the porous anodic alumina substrate. The films were tested via thermal cycling and gas separation experiments. The palladium membrane showed clear cracks after thermal cycling while the palladium silver alloy membranes showed no obvious damage. Gas separation tests showed that a membrane with 30% silver displayed increasing permeance (flux) with increasing temperature, with H_2/N_2 permselectivity of ~500 and permeance of ~1.8 $\mu mol/m^2.s.Pa$ at 700°C.

INTRODUCTION

        The increased uptake of sustainable hydrogen energy technologies is dependent on many technical, political and commercial factors, but a dominant technology issue is the provision of suitable quantities of high purity hydrogen. Low temperature fuel cells in particular are very sensitive to even trace levels of CO, SO_2, H_2S and other impurities in the supply gas stream, which can cumulatively damage and deactivate the electrode catalysts and reduce fuel cell efficiency[1,2]. Membrane filters provide a means to achieve very high purity hydrogen gas, especially through the use of palladium or palladium alloy materials that provide hydrogen-specific conduction paths[3,4]. Hydrogen diffusion through these membrane filters is a multistep process: dihydrogen molecules adsorb onto the surface of the Pd membrane and dissociate into hydrogen atoms, which then diffuse through the interstitial octahedral sites in the fcc Pd lattice and reform as gaseous diatomic hydrogen on the exit face of the palladium membrane. Palladium has a high level of hydrogen permeability whilst remaining impermeable to other gases, which makes it an ideal material for use to separate hydrogen from mixes of gases. Problems arise because palladium hydride exists in two different phases, α and β, both fcc structured as in the parent metal. When hydrogen is absorbed into Pd below 300 °C the high hydrogen capacity β hydride ($PdH_{0.67}$) predominates. Formation of this phase results in a 10% volume expansion causing stress on the Pd metal lattice and resulting in embrittlement and delamination of the material on repeated adsorption-desorption cycling[5,6]. However at temperatures above 300 °C the lower hydrogen capacity α hydride phase ($PdH_{0.02}$) predominates and the volume expansion upon adsorption of hydrogen is very small. As such, Pd films used in the presence of hydrogen are useful membrane materials only at temperatures above 300 °C.

        The introduction of other metals to form palladium based alloys has had promising results. In particular doping of the palladium with silver has been shown to improve the stability of the film and increase the solubility of hydrogen. Previous work by Uemiya et al.[7,8] showed that the temperature above which the α palladium hydride occurred was lowered with the increased silver content. The hydrogen permeability was optimized when the silver content of the alloy was around 23 wt%. Silver occupies interstitial sites in the palladium lattice and props them open to moderate the lattice expansion and contraction due to hydrogen absorption/desorption. The work reported here focuses on ways to control plating to obtain a Pd:Ag ratio of approximately 75:25 and measures the effect that varying the alloy silver content had on the hydrogen selectivity and permeability. This paper reports progress towards fabrication of ultrathin Pd and Pd-Ag films supported by robust and thermally stable ceramic substrates, suitable for hydrogen separation from gas mixtures at high temperatures.

Nanoscaled membranes offer specific advantages, including fast hydrogen transport via reduced diffusion path lengths through the metal and significantly reduced costs through reduction in the quantity of precious metal used. Our goal is to design a membrane to operate at temperatures above 500 °C to facilitate the separation of H_2 from hot synthesis gas streams derived from gasification plants. In previous work[9-14] we have shown that Porous Anodic Alumina (PAA) is a suitable template on which to support ultrathin Pd films. The ceramic template contains straight, parallel, uniformly sized pores of very high aspect ratio[15-17]. Its thermal stability depends on the specific anodization methodology used, particularly the selection of acid electrolyte. We have previously shown that PAA prepared in sulfuric acid is a suitable substrate material able to withstand >800 °C without deformation[9,11,13], whereas PAA commercially prepared in phosphoric acid shows substantial deformation above 700 °C[10-12] due to the asymmetric distribution of phosphate entities within the alumina ceramic matrix. Here we show how nanoscaled membranes can be fabricated by depositing ultrathin Pd and Pd-Ag films on durable ceramic templates fabricated using hard anodization techniques[13,14,17] and we demonstrate efficient hydrogen separation using these membranes.

METHODOLOGY

PAA membranes can be produced using both mild and hard anodizing techniques[9,18,19]. For mild anodization the voltage is maintained at a minimal level compatible with obtaining alumina growth. With an oxalic acid electrolyte, mild anodization is typically carried out at 40 V. Under these conditions the membrane growth occurs very slowly and it may take a period of several days to prepare a membrane of 100 μm thickness. The same thickness can be obtained using hard anodization techniques in about two hours. Hard anodization uses a much higher voltage to obtain higher current density and increased rate of alumina growth. Hard anodization using an oxalic acid electrolyte is carried out at 150 V. However if the anodization is commenced at 150 V the aluminum oxide grows unevenly resulting in an unusable membrane. Successful hard anodization requires formation of a stable template, which is normally a thin layer of porous alumina prepared under mild anodization conditions. Once this template is formed (around 10 minutes) the voltage can be safely increased to that selected for the hard anodization conditions. The square of aluminum was placed on a copper plate beneath the electrolytic cell and clamped in place. 0.3 M oxalic acid solution was used as the electrolyte. A platinum cathode was used and current was transferred to the aluminum anode via the copper plate. During anodization heat is produced which could lead to uneven pore growth leaving the membrane unsuitable for further use. To prevent the membrane from damage the cell was placed on a cooling plate which maintained a constant temperature of 3 °C throughout the anodization. Once the cell was assembled it was left for 20 minutes to equilibrate at 3 °C before the anodization was commenced. The voltage was maintained at 80 V for 10 minutes, forming a thin barrier layer upon which the remainder of the membrane grows. The voltage was then increased at 0.5 V/s to 150V. The thickness of the membrane is dependent on the amount of charge transferred. Higher voltage allows a greater rate of charge transfer; hence less time is required at a higher voltage to obtain the same thickness of PAA. The voltage was held at 150 V until 120 C of charge had been transferred (typically 12 h) after which the voltage was decreased at a rate of 0.05 V/s to 80 V. The 120 C corresponds to a PAA thickness of 120 μm. During the anodization, the electrolyte was stirred at a constant rate of 250 rpm using a magnetic stirrer. The cell design allowed the fabrication of 8 mm dia circular PAA discs, which retained their optically transparency even after subsequent detachment, pore widening and thermal procedures.

Fabrication and Detachment

Aluminum foil, 0.25 mm thick, 99.9995% pure (Alfa Aesar), was cut into 12 mm strips and cleaned by submersing in ethyl acetate and sonicating for 20 minutes. The cleaned foil was rinsed with distilled water and allowed to air dry. Once dry the strips of foil were cut to produce 12 x 12 mm squares. The electrolytic cell used for anodization is shown in Figure 1.

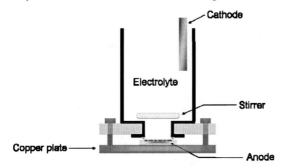

Figure 1. Schematic of Electrolytic Cell

The as-anodized PAA membrane remains attached to the aluminum substrate. To obtain the free-standing membranes used in these experiments the membrane was detached using a one-step voltage pulse detachment similar to that outlined by Yuan et al[20] and Chen et al.[21] This method has the advantages of being rapid and not requiring any heavy metals which may contaminate the membrane. Following anodization the oxalic acid electrolyte was discarded and after rinsing the cell with distilled water, was replaced with a 2:1 perchloric acid: ethanol solution. A voltage 10 V higher than the final anodization voltage was applied to the cell for 20 s. In this case the anodization concluded at 80 V so 90 V was used for detachment. After the single voltage pulse the perchloric acid-ethanol mixture was removed and the free-standing membrane was rinsed in distilled water. Membranes were stored in distilled deionized water until further use.

Pore Opening

Following detachment of the PAA from the aluminum substrate the pores of the PAA are closed at one end. Removal of this barrier layer opens the pores to form continuous collinear channels through the ceramic. This allows gas to flow freely through the ceramic during the gas separation testing, permitting measurement of the permeability of the applied metal coating. In addition to removing the barrier layer the pores may also widened to allow greater gas flow. These amorphous alumina membranes are susceptible to attack by acids and bases. Relatively mild conditions can be used to etch the membrane to remove the barrier layer and widen the pores. The small 80V "mild" section of the membrane is more susceptible to attack by acids than the main body of the membrane prepared under 150V "hard" conditions. By initiating the anodization at 80V the barrier layer is thus more susceptible to acid attack than the remainder of the membrane so the pore caps can be removed with little impact on the body of the membrane. Membrane etching was undertaken using a 5 wt% solution of phosphoric acid, placed in a beaker in a water bath and maintained at 30 °C. Both the temperature and time of immersion affected the final pore size and structure of the membrane.

Immersing the membranes for 60 minutes at 30 °C was determined as the ideal conditions to open the pores without damaging the structure of the membrane. The PAA membranes were rinsed in distilled water and allowed to air-dry.

Heat Treatment

As-anodized PAA membranes consist of amorphous aluminum oxide and are susceptible to attack at high and low pH conditions. This allows the successful pore opening of the membranes using a low concentration acidic solution. To ensure that the membranes could withstand exposure to highly basic plating solutions they were heat treated after pore opening to convert the amorphous alumina to γ-alumina. This was achieved by increasing the temperature at 10°C /min to 890 °C and holding for 5 minutes before cooling over four hours. γ-alumina is able to withstand the high pH plating solutions without damage to the chemical or physical structure of the membrane. Heat treatment to 890 °C also mitigates the risk of membrane cracking due to thermal expansion mismatch when curing the glass seal used for the gas test samples at 860 °C. We have previously reported extensive studies on the thermal and structural behaviour of PAA materials prepared using a range of acid electrolytes under both mild and hard anodizing conditions, characterized using XRD, thermal analysis and solid state ^{27}Al NMR techniques[9-14,18,19]. Species substituted in the alumina lattice from decomposition of the acid electrolyte play a major role in determining the chemical and physical stability of the ceramic template at elevated temperatures. The chemical structure of these ceramics develops from an X-ray amorphous state via one or more transition alumina phases, eventually crystallizing as α-Al_2O_3 (corundum) with the loss of bound carbon and sulfur species (as CO_2 or SO_2), depending on the acid electrolyte used. NMR shows that the amorphous structure contains a high degree of 5-coordinate Al bonding, while the nature of the high temperature transition alumina phases can be characterized using their 'AlO$_4$' and 'AlO$_6$' NMR signatures[9,12].

Metal Deposition

Pure palladium films undergo hydrogen embrittlement when maintained in a hydrogen containing atmosphere at temperatures below ~350 °C. The introduction of alloying metals such as silver has been shown to decrease the degree of hydrogen embrittlement and increase the hydrogen permeability of the coating. Palladium and silver containing films can be fabricated either by co-deposition or by individual deposition. In both cases the resulting film must be heat treated to alloy the silver and palladium.

A number of different methods can be used to plate PAA with metals. The electroless method used here was less arduous and more economical than the more typical electroplating methods. Electroless plating involves using a reducing agent to reduce metal ions in solution to the elemental metal. For electroless plating to be efficient and successful the surface of the membrane must be seeded with palladium crystals so that the reduced metal has a growth site[22]. If this activation is not undertaken the reduced metal will form on any available surface such as the sides of the container. Activation also ensures rapid growth of the metal coating, thus decreasing the time required for plating. This method of activation is commonly used to prepare surfaces for electroless plating[22,23]. The three solutions used for activation were a sensitizing solution consisting of $SnCl_2 \cdot 2H_2O$ (1 g L^{-1}) and HCl (33 %) (1 mL L^{-1}); an activation solution consisting of $PdCl_2$ (0.1 g L^{-1}) and HCl (33 %) (1 mL L^{-1}); and a solution of 0.01M HCl. The sensitizing solution was freshly made before use, since insoluble Sn(OH)Cl forms over time via the reaction scheme in equation (1). Hydrochloric acid was added to the solution to extend the amount of time for the precipitation to occur by increasing the concentration of the products.

$$SnCl_2 + H_2O \rightarrow Sn(OH)Cl + HCl \qquad \qquad \ldots\ldots(1)$$

The membranes were attached to a plastic holder using a small amount of double-sided tape. The membranes were immersed in solution one for five minutes, then rinsed with distilled water. This sensitization step deposits tin (II) hydroxide particles onto the membrane surface[22,23]. The sensitized membranes were then immersed in the activation solution for five minutes. During this step the tin (II) hydroxide particles reduce the Pd (II) to Pd metal via equation (2). The membrane was then briefly immersed in a dilute acid solution to prevent the hydrolysis of the freshly deposited palladium. The entire sensitization and activation process was repeated six times to obtain a sufficient coating of Pd to seed the electroless plating process.

$$Pd^{2+} + Sn^{2+} \rightarrow Pd^0 + Sn^{4+} \qquad \dots\dots (2)$$

An electroless plating method was used for all co-deposition and individual depositions. Electroless plating involves use of a plating solution containing a metal salt and stabilizing agents. A reducing agent was added just prior to plating which caused the metal salt to reduce to the elemental metal. Electroless plating is a widely used tool to deposit palladium[22,23] and palladium alloys[24,25] onto substrates. 25 mL of the plating solution was placed in a plastic cup in a water bath heated at 60 °C. Once equilibrated, hydrazine was added to the solution which was stirred briefly before the membrane was submerged in the solution. The samples were left for varying amounts of time depending on the desired thickness of the coating. The time was varied in the individual deposition to control the palladium to silver ratio of the coating. After plating was complete the samples were rinsed in distilled water and allowed to dry.

Palladium and silver can be co-deposited onto the PAA membranes using electroless plating solutions containing different ratios of Pd and Ag. Co-deposition has the advantage of interspersing the Pd and Ag particles resulting in increased intermetallic contact and more effective alloying[24,25]. A range of plating solutions were used composed of different Pd:Ag ratios. The metal salts used were $(NH_3)_4Pd(NO_3)_2$ and $AgNO_3$ and the amount used of each was varied to give solutions with different ratios of Pd and Ag whilst maintaining a metal concentration of 10mM. The overall composition of the solutions is shown in Table I. The components were added in the order given in the table. Ammonia was added to the water first then the Na_2EDTA. After the EDTA had dissolved, the palladium and silver salts were added. EDTA acted as a buffer to prevent the metal salts from precipitating. The hydrazine reduced the metal ions to metal and was added to the solution just prior to plating to prevent premature reduction of the metals.

Table I. Co-deposition Plating Solutions

	250 mL solution
H_2O	150 mL
NH_4OH (33 wt%)	94.2 mL
Na_2EDTA	15.0 g
Metal	10 mM
N_2H_4	2.5 mL

Individual deposition of the palladium and silver was also undertaken using an electroless method. By plating each metal individually the process can be controlled to obtain a range of different metal ratios. The solutions used are shown in Table II. The components of the solutions were the same as the co-deposition solutions except for palladium where a chloride salt was used in place of the nitrate. Membranes that were coated with palladium only were used to compare with the bimetallic sample. Samples were also plated with palladium then sequentially with silver as comparisons for the co-deposited samples.

Table II. Individual Plating Solutions

	Palladium	Silver
H_2O	250 mL	45 mL
NH_4OH (33 wt%)	2.5 mL	55 mL
Na_2EDTA	17.5 g	3.36 g
$Pd(NH_3)_4Cl_2$	0.4 g	0.054 g
$AgNO_3$	-	0.486 g
N_2H_4 (1M)	0.35 mL	0.3 mL

Alloying

Freshly plated samples contain both Pd and Ag phases which must be alloyed for the purpose of assessment of the influence of the addition of silver to the gas separation and thermal stability behaviour of the films. The alloying was carried out in an atmosphere controlled tube furnace under Ar or forming gas ($N_2/5\%H_2$), with a range of hold times between 4 -12 h and temperatures between 450-650 °C. The success of the heat treatment was determined by X-ray diffraction (XRD) (Bruker D8 Avance, CoKα radiation).

Thermal Cycling

A thermal cycling test was carried out to determine the relative stability of a pure Pd film and a Pd:Ag 75:25 co-deposited film. The PAA membranes were plated for 1 h, one with Pd only and one co-deposited with Pd:Ag 75:25 co-deposition solution. Following deposition, the bimetallic membrane was alloyed by heating to 500 °C for 12 h in forming gas. The Pd only and alloyed membranes were tested simultaneously. The membranes were placed in a tube furnace under 5% H_2 in N_2 gas. After gas equilibration in the chamber the furnace was ramped to 400 °C at 10 °C/min, the sample was held was 400 °C for 10 min. before cooling to 100 °C at 50 °C/h. The program was repeated to give a total of six cycles. The samples were inspected for differences using Scanning Electron Microscopy (JEOL 6500F).

Gas Separation

The samples used for gas separation required additional preparation. Following heat treatment at 900 °C the PAA membrane was mounted onto a ceramic ring, with a 5 mm central hole to allow the passage of gases. The PAA membrane was mounted over this ring using a glass seal, which was cured by heat treatment at 860 °C via ramping at 10 °C/min, holding at 860 °C for 10 min then cooling over 4 h. The heat treatment to 900 °C was carried out prior to mounting the membrane onto the disc. During this treatment the ceramic was converted from amorphous alumina to γ-alumina which occasionally resulted in slight curling of the membrane. This thermal pre-treatment was necessary to eliminate any risk of ceramic distortion which might otherwise result in cracking at the glass-ceramic interface due to thermal expansion or thermal mismatch during treatment at 860 °C. The glass sealant can be used up its softening temperature of ~860 °C. After the membrane was mounted onto the ceramic disc the activation and plating steps were carried out as normal. The ceramic disc was covered in tape during the activation and plating to restrict the metal coating to the PAA substrate. Gas separation tests were carried out using a rig shown schematically in Figure 2. An alumina ceramic support disc with the PAA mounted on it was placed between two copper rings inside a stainless steel rig. Copper rings were used because of the high malleability of copper which allowed the apparatus to be gas tight. Prior to the gas tests the copper rings were softened by heating to 900°C and were also polished to ensure a good contact surface. The two halves of the stainless steel rig were clamped together using six stainless steel cap screws. The copper rings and stainless steel bolts corrode over time so were replaced before each test. The entire rig was placed inside a tube furnace to control temperature.

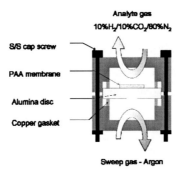

Figure 2. Schematic of Gas Separation Cell. PAA membrane is 8mm dia. for scale.

The analyte gas was a hydrogen-containing mixture consisting of 10% H_2, 10% CO_2 and 80% N_2. The pressure was increased on the analyte side of the membrane using a bubbler. The increased pressure assisted the diffusion of hydrogen through the membrane and encouraged diffusion in one direction only. Argon sweep gas was used on the other side of the membrane to collect gases that have diffused through the membrane. After flowing through the cell the sweep gas was analyzed online by mass spectrometry. Prior to the gas separation test the rig was leak tested to ensure integrity of the coating and sealant. Leaks through the membrane and glass seal do occur at a minimal level as indicated by the occasional observation of signals for mass 28 (N_2) and mass 44 (CO_2) in the mass spectrometry results.

Gas tests were carried out on co-deposited Pd:Ag 90:10, 80:20 and 75:25, a sequentially deposited sample and a palladium only sample summarised in Table III.

Table III. Gas Test Samples

Plating	Ratio Pd:Ag	Plating Time
Co-deposition	90:10	2 x 90 min.
Co-deposition	80:20	75 min. + 85 min.
Co-deposition	75:25	2 x 75 min.
Individual	Pd only	2 x 60 min.
Individual	78:22	Pd 75 min. Ag 2 x 20 min.

The gas test samples were put through a thermal cycling procedure. Once the mass spectrometer signals for the various gases had reached equilibrium at room temperature the temperature was increased to 600 °C over 24 h. The furnace was then cooled to 50 °C over 12 h before heating to 600 °C over 24 h and cooling to 50 °C again. The final cycle was from 50 °C to 700 °C. The Pd:Ag 90:10 sample was heated in 50 °C intervals allowing 30 min. at each temperature for the mass spectrometer signals to stabilize before taking a reading. This process was used to follow a similar heating schedule to 500 °C then 550 °C, returning to 50 °C by cooling between the heating cycles. The final ramp from 200 - 700 °C was undertaken using a steady heating rate for 36 h. The temperature schedule for the palladium only membrane was also an exception to this program. As noted earlier, palladium undergoes hydrogen embrittlement if heated in a hydrogen containing atmosphere below ~ 300 °C. Hence, the schedule for palladium involved heating to 350 °C under a nitrogen atmosphere then switching to the $10\%H_2/10\%CO_2/N_2$ analyte mixture. The mass spectrometer signals were allowed to settle overnight and the following day the sample was heated between 350 and 550°C over 8 h. The sample was then cooled to 350 °C over 4 h and heated back to 550 °C. The sample was finally cooled to 100 °C still under the hydrogen containing atmosphere.

RESULTS AND DISCUSSION
PAA Membrane Formation

SEM images of the basal and pore sides of a detached PAA membrane are shown in Figure 3. The majority of the barrier layer was not removed during detachment so many pores are still closed and those that are open are very narrow. For the membrane to be used effectively for gas separation the barrier layer must be entirely removed and the pores widened.

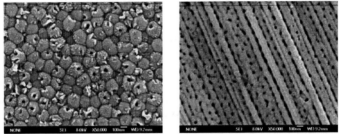

Figure 3. Basal (left) and pore (right) sides of a detached membrane. Scale bars are 100 nm.

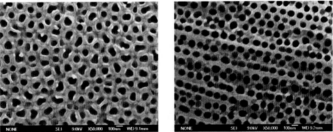

Figure 4. Basal (left) and pore (right) sides after pore opening in 5 wt% H_3PO_4. Scale bars are 100 nm.

Removal of the barrier layer and widening of the pores was carried out by immersion of the detached membranes in a 5 wt% phosphoric acid bath at 30 °C for 1 h. The phosphoric acid etched the alumina slowly, dissolving it in a manner similar to the perchloric acid during detachment. However the phosphoric acid was in contact with the PAA for a longer time allowing the dissolution of the pore caps and widening of the pores, as can be seen in Figure 4. Up to four membranes were pore opened at a time, with the acid bath replaced after every batch.

Metal Deposition on PAA Membranes

Activation is used to seed the surface of the PAA membrane with palladium grains to form sites for growth of the palladium coating during electroless plating. The sensitization step deposits tin hydroxide particles on the surface of the PAA. During immersion in the activation solution the tin hydroxide particles reduce the palladium to palladium metal which then deposits on the surface of the membrane. No tin hydroxide residue remains on the surface of the PAA after activation. The surface of an activated PAA membrane is shown in Figure 5.

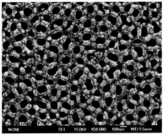

Figure 5. Activated PAA membrane. Scale bar is 100 nm.

Co-deposition of Pd and Ag is desirable as it creates a layer of interspersed Pd and Ag grains, which aids alloying. A ratio of Pd:Ag 75:25 was sought based on previous work[7] which reported that an alloy content of 23% Ag had the most selective hydrogen permeability, and the lowest critical temperature. Experiments were carried out using a number of plating solutions with different Pd:Ag ratios. All solutions gave an even coverage of Pd and Ag. Membranes plated in Pd:Ag 75:25 and 50:50 solutions appeared as a silvery white. Those coated with 90:10 appeared to be more silvery in colour. EDX analyses carried out on samples plated for 30 minutes in solutions containing Pd:Ag 75:25 and Pd:Ag 50:50 showed that the films had a higher Ag content than that of their parent solution; eg. for the 75:25 solution the composition of the coating after 60 minutes was 63:37. This phenomenon has been previously reported[24] and indicates that Ag is preferentially plated resulting in Ag enrichment. A Pd:Ag 90:10 solution examined using EDX analysis showed that a 90:10 film co-deposited for 60 minutes resulted in a Pd:Ag ratio of 73:27. The inconsistency between the Pd:Ag ratio in the solution compared to the resulting film encouraged us to experiment with individual deposition. The 75:25 plated membranes showed another interesting trend. EDX mapping of the Pd and Ag atoms in the film was performed on membranes plated for 30 and 60 minutes. Both samples showed nodes on the surface. These were confirmed to contain mostly Ag whilst the area around the nodes contained mostly Pd. In addition the EDX results showed the relative amount of Pd to increase between the 30 to 60 min depositions. These results suggest that Ag is deposited first in nodes on the surface followed by a slower build-up of Pd in the area around the nodes. Hence modifying the duration of plating will affect the ratio of Pd to Ag in the film.

Individual deposition was attempted as a means to produce films with more controllable Pd and Ag ratios. The individual deposition requires activation prior to deposition as with the co-deposition solutions. Initially only two layers were attempted, the first Pd and the second Ag. This was successful and the ratio of Pd and Ag could be easily controlled. However the resulting film was very thin and thus unsuitable for a gas separation membrane as it was more prone to leaks and degradation. Preparation of a thicker film was attempted by plating Pd then Ag then Pd. This sandwich structure should promote alloying by increasing the inter-metallic surface area. However the second Pd deposition resulted in minimal weight increase indicating that deposition had not occurred. The initial Pd deposition requires a Pd-seeded surface, so it is likely that the lack of such a surface has prevented the second Pd deposition from occurring. The membranes were dried and weighed before and after each plating step to gain accurate knowledge of the deposition vs time process and enable control of Pd:Ag composition. This data is not reproduced here. We observed that fresh solutions gave more reproducible results than aged solutions, so all gas test samples were prepared using freshly made solutions. The surface of the pure Pd film after coating for 30 min. is shown in Figure 6. The surface is uniform and even, but still shows some evidence of pinholes in the structure. Plating for longer than 30 min. decreased the likelihood of pinholes and other defects.

Figure 6: Surface of a Palladium only film after 30 min. deposition. Scale bar is 1 micron.

Alloying

Thermogravimetry and differential scanning calorimetry experiments (not reproduced here) were carried out on a range of samples in air, N_2 and Ar. The results were similar between atmospheres and the Pd:Ag 75:25 and 50:50 results were similar. The bimetallic samples underwent exothermic events at ~350 and 750 °C, whilst the Pd only and uncoated samples only showed exothermic events at 650 °C. All thermal events were associated with a loss in mass. It is proposed that the transformation at 350 °C is due to annealing of the metals as this exotherm was only apparent in the metal coated samples. From this data, alloying was undertaken at 450 °C for 4 h, initially in argon. However, alloying did not occur under these circumstances. Although 450 °C was sufficient for alloying after holding for 12 h, the film delaminated making it unusable for gas separation. This delamination also occurred at 550 °C and 650 °C and appeared to be a result of using argon. When alloying was undertaken in forming gas (5% H_2 in 95% N_2) no delamination was evident. Under forming gas the conditions were refined, showing that 500 °C for 12 h was sufficient for alloying. The success of the heat treatment was determined by XRD as shown in Figure 7.

The 'as-plated' scan of a pre-alloyed sample shows two peaks, at approx. 45° 2θ and 47° 2θ due to Ag and Pd respectively. The 'after heat treatment' scan shows successful alloying revealed by a single peak at approx. 45.5° 2 θ. The position of the alloyed peak can be used as a tool to estimate the relative proportions of Pd and Ag in the alloy, as demonstrated in the XRD patterns in Figure 8.

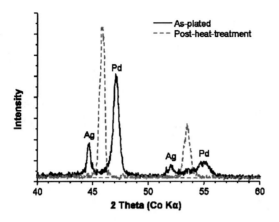

Figure 7. XRD scans before and after alloying.

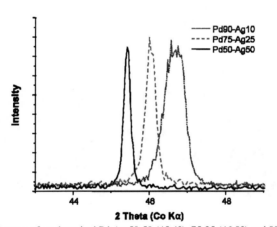

Figure 8. XRD scans of co-deposited Pd:Ag 50:50 (45.4°), 75:25 (46.0°) and 90:10 (46.7° 2θ).

The surfaces of two different samples after alloying are shown in the SEM images in Figure 9. These samples were heat treated under Ar to 550 °C for 12 h. The surface of the membranes show islands of alloyed metal surrounding areas where the metal is no longer present. This indicates that whilst the metal covered the PAA surface completely after plating, this amount of metal is insufficient to cover the surface after alloying. As a result the gas test samples were plated for multiple and longer times to ensure complete coverage after alloying. The desired coverage was achieved by lengthening the plating time to 90 min. as can be seen in the Pd:Ag 90:10 sample (Fig. 9, right).

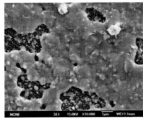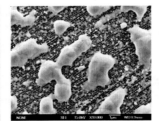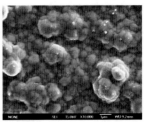

Figure 9. SEM images of membrane surfaces after alloying.
Left: Pd:Ag 50:50, 30 min; *Centre*: 75:25, 30 min; *Right*: 90:10, 60 min.

Thermal Cycling Tests

A palladium only membrane and a membrane coated using Pd:Ag 75:25 co-deposition and alloying, were exposed to thermal cycling testing from 100 to 400 °C. The heating rate was 600K/hr and cooling was 50K/hr. The process was repeated to give a total of six heating and cooling cycles. After the thermal cycling no change in the membranes quality could be detected by eye. However the SEM results shown in Figure 10 indicate a substantial difference between the membranes. The Pd only membrane (initially the same material as in Figure 6) has developed large cracks across the surface, whilst the bimetallic sample has no obvious defects. Before cyclic testing, the Pd:Ag, 75:25 co-deposited membrane had large nodes of silver on the surface (Fig 10, centre). The cyclic test has caused these to become less crystalline and more rounded in appearance and it is possible that over time this change may cause defects in the membrane; however the addition of silver to the membrane has increased the thermal stability and as a result the life of the membrane.

Figure10. SEM images of membrane surfaces after thermal cycling. *Left*: Pd only, after cycling (cf. Fig 6); *Centre*: Pd:Ag 75:25, before cycling; *Right*: Pd:Ag 75:25, after cycling.

Gas Separation Tests

Gas separation tests involved heating the samples at a constant rate while exposed to a hydrogen-containing gas mix (10% H_2/10% CO_2/80% N_2) and analyzing the composition of the Ar sweep gas using mass spectrometry. If H_2 (or other gases) had flowed through the membrane they would be picked up by the sweep gas and their signals seen by the mass spectrometer. The rig used is not ideal and some analyte gas was able to enter the sweep gas by flowing past the ceramic support disc. As such a baseline level for H_2, CO_2 and N_2 is present in the data. All of the Ag containing samples were exposed to H_2 at low temperatures which did not appear to cause any damage to the membrane as no significant change in either N_2 or CO_2 levels occurred. If the membrane had been damaged these gases would be able to flow through the membrane causing an increase in the N_2 or CO_2 mass signals. The Pd only results shown in Figure 11 have no data between room temperature and 350 °C as the membrane was heated under a H_2 free atmosphere in this temperature window to prevent embrittlement. Above 350 °C the H_2 flux (permeance) increases significantly whilst the N_2 and CO_2 fluxes remain at a minimal level.

The Pd:Ag 90:10 co-deposited specimen was sealed effectively with minimal leaks. Though there is some variability in the data (plotted in Figure 12), overall the H_2 flux increases with increasing temperature, while the flux of both N_2 and CO_2 remain stable at a minimal level, essentially at the detection limit of the mass spectrometer. Up to ~ 350 °C the β hydride phase will dominate and our results show that above 350 °C there is a more rapid increase in H_2 flux through the membrane with accelerated transport through the α phase. This increase in H_2 flux above 350 °C is consistent with the results for the Pd only membrane (Figure 11) indicating that increasing the Ag content to ~ 30% (by EDX) has no detrimental effect on the H_2 selectivity.

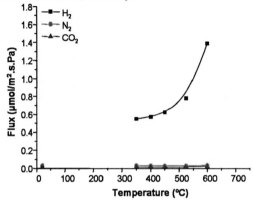

Figure 11. Gas Separation Results for Palladium only Membrane

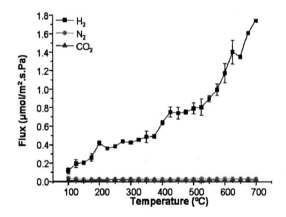

Figure 12. Gas Separation Results for Pd:Ag 90:10 Membrane

Both the Pd:Ag 80:20 and 75:25 co-deposited samples did not seal as effectively as the 90:10 sample. These samples had higher baseline levels of both CO_2 and N_2. Although the data is not reproduced here the CO_2 and N_2 remain stable throughout the test indicating that the leak is due to a failure of the rig rather than a flaw in the membrane. The change in H_2 permeance is much less dramatic than for the 90:10 sample however the measured permeance does increase steadily with temperature to ~1.0 $\mu mol/m^2.s.Pa$ at 700 °C, compared to with ~1.8 $\mu mol/m^2.s.Pa$ for the 90:10 membrane (Figure 12). This most likely a result of the increased Ag content in the Pd:Ag 80:20 and 75:25 materials. An "ideal" Ag content of ~23wt% has been shown to display the highest H_2 selectivity and thermal stability[7]. The actual Ag content in both of our 80:20 and 75:25 films is higher than this 23wt% ideal and as such the films will be more thermally stable than those with higher Pd content. However, these higher Ag content alloys should also display reduced H_2 permeance, as supported by our data. EDX analysis indicates that the composition of our "90:10" film was actually close to the desired 23wt% Ag content so the H_2 permeance might be expected to increase compared to the Pd only membrane, although our data at 600 °C (Figs 11 &12) show little distinction in flux.

The method used for these gas separation tests involved several cycles of heating and cooling, as previously noted. Each of these membranes shows a degree of thermal stability over the three cycles as no increase in nitrogen or carbon dioxide was observed indicating no damage to the film. The results discussed here are those for the third (final) cycle of each of the tests and it is instructive to note that the flux increased considerably between cycles, indicating that greater benefit may well be obtained following more extended thermal cycling. This is under examination at present.

CONCLUSIONS

Porous Anodic Alumina (PAA) ceramic templates have been fabricated by applying a combination of mild and hard anodizing conditions to high purity Al foils. Following electrochemical detachment and acid pore opening procedures, electroless deposition techniques were used to apply ultrathin coatings of palladium and silver metal. Co-deposition of Pd and Ag is better than individual deposition as the metal grains are more homogeneously interspersed which results in more effective alloying. However further work is required to better manage the co-deposition process, to overcome our consistent observation that the deposited film had a greater silver content than that of the plating solution used. Co-deposition is also advantageous, since with using individual deposition we were unable to coat a silver film with palladium so a sandwich structure could not be used for individual deposition.

Suitable conditions for alloying the Pd-Ag were determined as 500 °C, for 12 h. under forming gas (5% H_2 in N_2). In addition the position of the XRD alloying peak can be used as a good guide to the relative composition of a film and with suitable calibration this would be a very accurate guide.

Thermal cycling experiments showed that the Pd-Ag alloyed membrane did not undergo hydrogen embrittlement and was more thermally stable than a Pd only film which developed large cracks. This improvement in thermal stability was also observed in gas separation tests as the co-deposited samples were able to maintain selectivity after several days of exposure to hydrogen including at temperatures below 350 °C, where the Pd-only membrane failed. The gas tests showed that the hydrogen flux increased with temperature in all cases, however the Pd:Ag 90:10 was the best of the tested membranes as it had the greatest increase in H_2 flux whilst maintaining high hydrogen selectivity. The 80:20 and 75:25 samples had higher baseline levels of N_2 and CO_2, possibly due to a leak in the system, but it also may be due to the high silver concentrations of these membranes which may reduce their selectivity towards hydrogen.

Overall we conclude that the addition of silver to a palladium membrane resulted in increased stability when exposed to hydrogen at low temperatures and that the ideal silver concentration is around 30% (as obtained from the 90:10 co-deposition solution). This membrane displayed the greatest hydrogen permselectivity (H_2/N_2 ~500) during gas separation tests.

ACKNOWLEDGEMENTS
The authors wish to thank David Flynn, SCPS Victoria University of Wellington for excellent support with SEM imaging. Funding for this research was provided by the MacDiarmid Institute for Advanced Materials and Nanotechnology and by the New Zealand Foundation for Research Science and Technology (FRST) via contract number CO8X0706.

REFERENCES
1. Francisco A. Uribe, Judith A. Valerio, Fernando H. Garzon, and Thomas A. Zawodzinski, PEMFC Reconfigured Anodes for Enhancing CO Tolerance with Air Bleed, Electrochem. Solid-State Lett. **7** A376 (2004).

2. J.J. Baschuk and Xianguo Li, Modelling CO Poisoning and O_2 Bleeding in a PEM Fuel Cell Anode, Int. J. Energy Res. **27**, 1095–1116 (2003).

3. S. N. Paglieri, Palladium Membranes, in Nonporous Inorganic Membranes. Published Online: 3 Aug 2006, Eds. A.F. Sammells, M.V. Mundschau, Print ISBN: 9783527313426, Online ISBN: 9783527608799, pp77-105.

4. T. B. Flanagan and Y Sakamoto, Hydrogen in Disordered and Ordered Palladium Alloys, Platinum Metals Rev., **37(1)**, 26-37 (1993).

5. S H Goods and S E Guthrie, Mechanical properties of palladium and palladium hydride Scripta Metallurgica **26(4)**, 561-566 (1992).

6. V V Bhat, C I Contescu and N C Gallego, The role of destabilization of palladium hydride in the hydrogen uptake of Pd-containing activated carbons, Nanotechnology **20** (2009) 204011 (doi:10.1088/0957-4484/20/20/204011).

7. E. Kikuchi and S. Uemiya, Preparation of supported thin palladium-silver alloy membranes and their characteristics for hydrogen separation, Gas Separation & Purification, **5**, 261 (1991).

8. S. Uemiya, T. Matsuda and E. Kikuchi, Hydrogen permeable palladium-silver alloy membrane supported on porous ceramics, Journal of Membrane Science, **56**, 315-325 (1991).

9. A. Kirchner, K.J.D. MacKenzie, I.W.M. Brown, T. Kemmitt and M.E. Bowden, Structural characterisation of heat-treated anodic alumina membranes prepared using a simplified fabrication process, Journal of Membrane Science **287**, 264-270 (2007).

10. I.W.M. Brown, M.E. Bowden, T. Kemmitt and K.J.D. MacKenzie, Structural and Thermal Characterisation of Nanostructured Alumina Templates, Curr. Appl. Physics **6**, 557-561 (2006).

11. A. Kirchner, I.W.M. Brown, M.E. Bowden, T. Kemmitt and G. Smith, Preparation and high-temperature characterisation of nanostructured alumina ceramic membranes for high value gas purification, Curr. Appl. Physics **8**, 451-454 (2008).

12. I.W.M. Brown, M.E. Bowden, T. Kemmitt, A. Kirchner and K.J.D. MacKenzie, New Ceramic Membrane Materials for Gas Purification, Adv. Materials Research **29-30**, 15-20 (2007).

13. A. Kirchner, I.W.M. Brown, M.E. Bowden and T. Kemmitt, Hydrogen Purification using Ultra-thin Palladium Films supported on Porous Anodic Alumina Membranes, in Functional Nanoscale Ceramics for Energy Systems, ed. E. Ivers-Tiffee and S. Barnett (Mater. Res. Soc. Symp. Proc. 1023E, Warrendale, PA, 2007), Paper 1023-JJ09-02.

14. I.W.M. Brown, M.E. Bowden, E. Jay, T. Kemmitt, A. Kirchner, K.J.D. MacKenzie & G. Smith, in Global Roadmap for Ceramics – Proceedings of ICC2, (ISTEC-CNR, Institute of Science and Technology for Ceramics - National Research Council, 2008) p319-328.

15. N. Itoh, N. Tomura, T. Tsuji and M. Hongo, Deposition of palladium inside straight mesopores of anodic alumina tube and its hydrogen permeability, Microporous Mesoporous Mater. **39**, 103-111 (2000).

16. R. Inguanta, M. Amodeo, F. díAgostino, M. Volpe, S. Piazza and C. Sunseri, Developing a procedure to optimize electroless deposition of thin palladium layer on anodic alumina membranes, Desalination **199**, 352-354 (2006).

17. W. Lee, R. Ji, U. Gösele and K. Nielsch, Fast fabrication of long-range ordered porous alumina membranes by hard anodization, Nature Materials 5, 741-747 (2006).

18. A. Kirchner, I.W.M. Brown, M.E. Bowden, T. Kemmitt, G. Smith, Preparation and high-temperature characterisation of nanostructured alumina ceramic membranes for high value gas purification, Current Applied Physics 8, 451-454 (2008).

19. Ian Brown, Mark Bowden, Tim Kemmitt, Jeremy Wu and Jules Carvalho, Nanostructured Alumina Ceramic Membranes for Gas Separation, International Journal of Modern Physics B. 23(6), 1015-1020 (2009).

20. J.H. Yuan, W. Chen, R.J. Hui, Y.L. Hu and X.H. Xia, Mechanism of one-step voltage pulse detachment of porous anodic alumina membranes, Electrochimica Acta 51, 4589–4595 (2006).

21. W. Chen, J. S. Wu, J. H. Yuan, X. H. Xia, X. H. Lin, An environment-friendly electrochemical detachment method for porous anodic alumina, Journal of Electroanalytical Chemistry, 600, 257-264 (2007).

22. P. P. Mardilovich, Y. She, Y. H. Ma, Defect-free palladium membranes on porous stainless steel support, AIChE Journal, 44(2), 310 (1998).

23. J. P. Collins, J. D. Way, Preparation and Characterization of Palladium-Ceramic Composite Membranes, Ind. Eng. Chem, 32, 3006-13 (1993).

24. Y.S. Cheng, K.L Yeung, Palladium-silver composite membranes by electroless plating technique. Journal of Membrane Science 158, 127-141 (1999).

SOFTENING OF RARE EARTH ORTHOPHOSPHATES BY TRANSFORMATION PLASTICITY: POSSIBLE APPLICATIONS TO FIBER-MATRIX INTERPHASES IN CERAMIC COMPOSITES

R. S. Hay, G. Fair
Air Force Research Laboratory
Materials and Manufacturing Directorate
WPAFB, OH

E. E. Boakye, P. Mogilevsky, T. A. Parthasarathy
Air Force Research Laboratory
UES, Inc., Dayton, OH

J. Davis
Wright State University
Fairborn, OH

ABSTRACT

Rare-earth orthophosphate interphases made from nanoparticle precursors have been successfully demonstrated for dense matrix oxide-oxide CMCs. For these interphases the major concern is high fiber pull-out stresses, typically ~80 - 200 MPa. Plastic deformation mechanisms in a 10 – 100 nm thick zone of rare-earth orthophosphate adjacent to the fiber govern pullout friction. For lower fiber pull-out stresses, rare-earth orthophosphates and vanadates that soften by transformation plasticity during the martensitic xenotime $\rightarrow$ monazite phase transformation were investigated. Predictive methods developed for prediction of deformation twinning in orthophosphates were extended to transformation plasticity. Nano-indentation testing was used to develop and test materials suitable for transformation plasticity weakened fiber coatings. Transformation plasticity significantly softens $TbPO_4$ and $(Gd,Dy)PO_4$ solid-solutions in the xenotime phase. Transformed regions were characterized by TEM; some evidence suggests that the phase transformation may reverse with time. Preliminary attempts to coat single-crystal alumina fibers with these materials were made. The potential to tailor fiber-matrix interphase friction in CMCs is discussed.

INTRODUCTION

Rare earth orthophosphates are stable at high temperatures in oxidizing and high vacuum environments.[1-4] Oxide-oxide CMC's with rare-earth orthophosphate interphases such as monazite and xenotime have been successfully demonstrated by a number of different research groups.[5-10] However, these materials are much less mature than other CMCs. The available information suggests that performance limiting factors for oxide-oxide CMCs are: 1) crack deflection and fiber-pullout shear stress (friction) of the rare-earth orthophosphate fiber-matrix interfaces,[11] 2) creep and high temperature strength of the nano-grain size (60 – 100 nm) oxide fibers, particularly in humid or combustion environments,[12,13] and 3) environmental effects, particularly environmentally assisted subcritical crack growth (EASCG).[14-17]

The major concern for rare-earth orthophosphate interphases is the high fiber pull-out stresses, typically ~80 - 200 MPa, which may be near the borderline of acceptable values.[1,18,19] These stresses contrast with 5 – 20 MPa typically measured or inferred for carbon and BN interphases.[20,21] High pullout stress biases fiber failure towards shorter pullout lengths, which may slightly increase strength, but at the expense of toughness (strain to failure) and flaw tolerance, the most desirable CMC attributes.[22,23] The optimal pull-out stresses are not known, and depend on the particular structural application. Pullout friction is governed by the plastic deformation mechanisms of a thin layer (10 – 100 nm) of the interphase material adjacent to the fiber (Fig. 1).[24] These mechanisms include dislocation slip, microfracture, formation of deformation twin nano-lamellae, and cataclastic flow of deformed nanoparticles (Fig. 1).

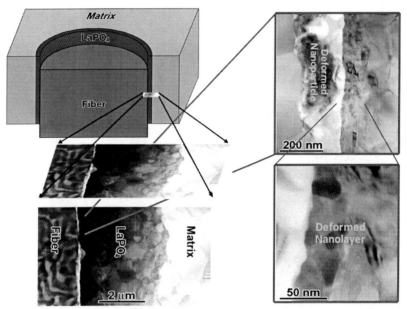

Fig. 1 *TEM image sequence showing deformation in the monazite fiber-matrix interphase during room-temperature fiber pushout.*[24] *Deformation is concentrated in a ~100 nm thick layer adjacent to the fiber. An intensely deformed cataclastic nanoparticle and a deformed, recrystallized layer are marked.*

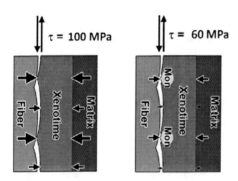

Fig. 2 *An illustration of idealized transformation plasticity weakening by a xenotime coating on a fiber. Local high pressures and shear stress from fiber roughness trigger the xenotime → monazite phase transformation,n which weakens the interphase material and lowers push-out friction.*

Some approaches to lowering pullout stress are use of softer ABO_4 phases such as tungstates, vanadates, or niobates, or nanoscale texturing of such materials so that cleavage or slip planes are in the plane of the fiber-matrix interface. However, many of these phases are unlikely to be stable with CMC constituents at high temperatures.[25] Another approach to lowering pullout stress involves use of interphases that undergo a $-\Delta V$ martensitic phase transformation. The concept is illustrated for the xenotime → monazite phase transformation in figure 2. The phase transformation can be driven by local high pressures and shear stresses caused by accommodation of fiber roughness during pullout. There are two possible effects; friction reduction by transformation plasticity, or by local reduction of normal stress from contraction of small volumes of the interphase material during the phase transformation (Fig. 2).

TRANSFORMATION PLASTICITY

Transformation plasticity occurs when the atomic rearrangements during a phase transformation simultaneously accommodate stress and therefore weaken the material. It has been known to metallurgists for over 80 years,[26-30] but the phenomena has seen little application in ceramics (a brittle-field mechanism, called "transformation weakening", was proposed for interphase crack deflection,[31,32] but not for reduction of pullout friction). Transformation plasticity has also been proposed to weaken earth materials.[28,33]

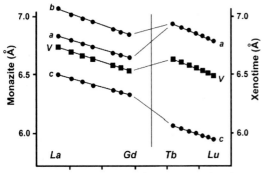

Fig. 3. *Lattice parameter of the rare earth phosphates. The stable phase changes from monazite to xenotime between GdPO₄ and TbPO₄. The xenotime phase has ~5.8% larger molar volume than the monazite phase.*

Large rare-earths, such as La and Ce, form orthophosphates with the monazite structure. Small rare earths, such as Y and Lu, form orthophosphates with the xenotime structure. The change between the structures occurs between the $GdPO_4$ and $TbPO_4$ compositions (Fig. 3, 4).[34] Similar observations have been made for rare-earth orthovanadates; here the change from monazite to xenotime structures occurs between $LaVO_4$ and $CeVO_4$.[35] The transformation between the two structures should be martensitic and kinetically facile, since only small atomic shuffles, half the magnitude of those involved in (100) and (001) monazite deformation twinning, are involved (Fig. 5).[36] The shear accompanying the xenotime → monazite transformation does not have a shear sense, or sign, unlike deformation twinning,[37] and because of the tetragonal xenotime symmetry, the transformation can occur on (100), (010), and (001) in xenotime. Pressures that induce the transformation can be estimated from thermodynamic data,[38-42] and are ~ 1 GPa for $TbPO_4$ (Fig. 6), but approach zero for $(Dy_x,Gd_{1-x})PO_4$ solid-solutions near monazite-xenotime equilibrium. Thermodynamic calculations suggests monazite-xenotime equilibrium for a $(Dy_{0.8}Gd_{0.2})PO_4$ solid-solution, with little temperature dependence.[43,44] It should be possible to shift the monazite-xenotime equilibrium to larger rare-earth cations, which are softer because

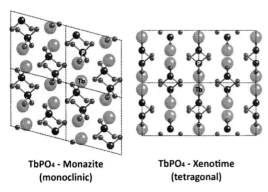

TbPO₄ - Monazite
(monoclinic)

TbPO₄ - Xenotime
(tetragonal)

Fig. 4. *Crystal structures for monazite and xenotime TbPO₄ polymorphs.*

of longer rare-earth – oxygen bonds, by making $RE(P_xV_{1-x})O_4$ solid-solutions. The trade-off for orthovanadates is that these compounds are less refractory and less stable, particularly with respect to reduction,

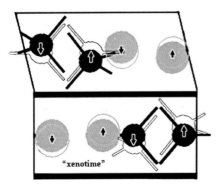

Fig. 5 *Shuffles for rare-earth cations (gray) and PO_4 tetrahedra during deformation twinning on the (100) plane in monazite. The maximum shuffle required for twinning is 0.07 nm, which is less than ½ the La-O bond distance so diffusion is not required. The atoms shuffle through the "xenotime" structure, which requires only 0.035 nm shuffles for a martensitic transformation.*

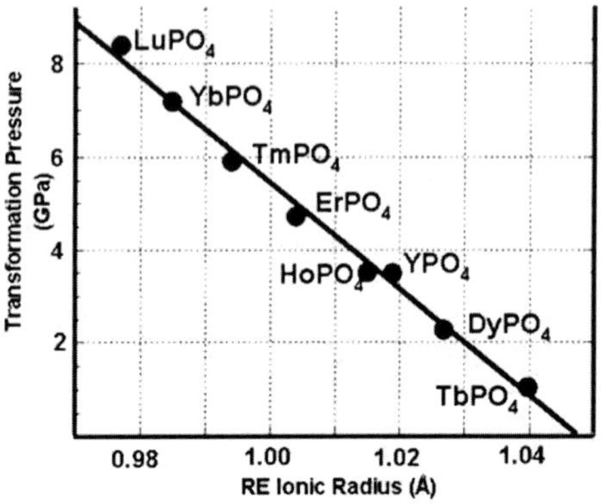

Fig. 6 *Calculated xenotime → monazite transformation pressures at room temperature for different rare-earth orthophosphates.*

in comparison with orthophosphates.[42]   The volume loss of the xenotime → monazite transformation (~5.8 vol%) is equivalent to ~2% linear shrinkage, and should in principle relax normal stress induced by interface roughness of 5 nm magnitude, and therefore the pullout friction, if the transformation occurs through an interphase 250 nm thick. The actual reduction in friction will depend on nanostructural details, such as the extent of and mechanisms by which the phase transformation is accommodated in the interphase. The ease of inducing the transformation, and the subsequent effect on mechanical properties through transformation plasticity, can be assessed by the stress-strain signature and hysteresis during instrumented nano-indentation, using a method that is widely applied to silicon.[45-48]

We present and discuss preliminary experiments on rare-earth orthophosphates that soften by transformation plasticity during the xenotime → monazite phase transformation. The effort involves identification of soft rare-earth orthophosphate compositions and solid-solutions, characterization of mechanical properties by instrumented indentation, and characterization/verification of fiber push-out/pullout friction of fibers coated with the identified compositions.

EXPERIMENTS
Rare-earth orthophosphate powders with $GdPO_4$, $TbPO_4$, and $DyPO_4$ compositions were prepared and characterized for phase and particle morphology by methods used previously.[49,50] A series of powders with $(Gd_x,Dy_{1-x})PO_4$ solid-solutions were also prepared. The powders were cold-pressed and sintered at 1600°C for 20 hours. Phase presence was checked using X-ray. The sintered pellets were sectioned and polished with diamond laps to < 1 μm surface finish. Indentation of the polished pellets was done using a MTS Nano Indenter XP system with Berkovich indenter, a 10 μm tip, and a 1:7 depth-to-side ratio. The indentation depth limit was 1.5 μm. Indentations were characterized by optical microscopy (reflected light) and SEM. Focused ion beam (FIB) sections were milled out beneath some $TbPO_4$ indentations and thinned to electron transparency. These sections were characterized in a Phillips CM200 FEG TEM operating at 200 kV.

Preliminary fiber coating experiments were conducted using $TbPO_4$ and $(Gd,Dy)PO_4$ solid-solution precursors described elsewhere.[50] Single-crystal alumina fibers (Saphikon™) were coated using heterogenous nucleation and growth from rare-earth citrates; the methods are described in other publications.[50-53] TEM specimens were prepared of the coated fibers by published methods. The coatings were characterized for microstructure, phase and composition in a Phillips CM200 FEG TEM operating at 200 kV using EDS and selected area electron diffraction. Coating thicknesses of 300 – 400 nm were used with a final coating heat-treatment of 5 minutes at 900°C. Coated fibers were slip cast in Sumitomo alumina powder and hot-pressed at 1400°C, 15 MPa for 1 hour to make minicomposites.

RESULTS AND DISCUSSION

Abnormal grain growth hindered densification of the $TbPO_4$ and $(Gd,Dy)PO_4$ xenotime pellets. Polishing of $TbPO_4$ and $(Gd,Dy)PO_4$ pellets with xenotime compositions near the monazite-xenotime phase boundary was also very difficult (Fig. 7). These material removal rate was significantly faster for these compositions than $GdPO_4$ and $DyPO_4$ compositions. This was attributed to poor densification, transformation plasticity, and to grain pullout driven by stress generated from local xenotime → monazite phase transformations during polishing. Pellets with $GdPO_4$ and $DyPO_4$ compositions were not difficult to polish.

Indentation experiments were done on the sporadically dispersed grains or clumps of grains that were well polished (Fig. 7). Indentation load-displacement curves for $GdPO_4$ (monazite), $DyPO_4$ (xenotime), and $TbPO_4$ (xenotime) are shown in figure 8. $GdPO_4$ is the rare-earth orthophosphate monazite phase closest to the xenotime stability field (Fig. 3). $TbPO_4$ is the rare-earth orthophosphate xenotime phase closest to the monazite stability field, and $DyPO_4$ is the xenotime phase that is next closest to the monazite stability field (Fig. 3).

Indentation load-displacement curves for $GdPO_4$ (monazite), $DyPO_4$ (xenotime), and $TbPO_4$ (xenotime) are shown in figure 8. The $TbPO_4$ was much softer than either $DyPO_4$ or $GdPO_4$. This is consistent with the presence of transformation plasticity in $TbPO_4$, but not in $DyPO_4$. There was much more variation in load for a 1500 nm displacement for different indentations in the $TbPO_4$ than either $DyPO_4$ or $GdPO_4$, suggesting significant anisotropy for transformation plasticity. Grain orientations with maximum shear stress along $<100>\{010\}$ and $[010](001)$ are expected to deform most easily by transformation plasticity (Fig. 5).

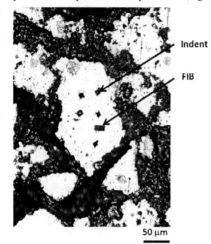

Indent

FIB

50 μm

Fig. 7 Optical micrograph (reflected light) of a polished TbPO₄ surface. Indentations, and the location of a FIB section through an indentation are marked.

The indentation load-displacement curves for $(Gd_{0.5},Dy_{0.5})PO_4$ (xenotime) are shown with the $DyPO_4$ (xenotime) load-displacement curve from figure 8 in figure 9. The $(Gd_{0.5},Dy_{0.5})PO_4$ is even softer than $TbPO_4$, and there were even larger variations in load for a 1500 nm displacement for different indentations. Some load-displacement curves exhibit up to 75% rebound hysteresis. This is diagnostic of a ferroelastic or shape memory material, and may be due to either phase transformation reversal or elastic deformation twinning.[54]

Several focused-ion beam (FIB) sections were cut directly beneath the bottoms of indentations in $TbPO_4$. Most show only intense plastic deformation upon examination by TEM. A couple did show conclusive evidence for the xenotime → monazite phase transformation. The best example is shown in

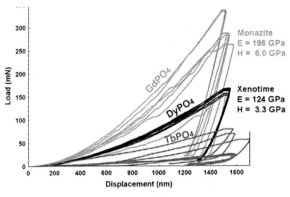

Fig. 8. *Indentation – displacement curves for GdPO₄ (monazite), DyPO₄ (xenotime),and TbPO₄ (xenotime). TbPO₄ is by far the softest material. Variation in load displacement curves may reflect anisotropy of transformation plasticity mechanisms related to the crystallography of the martensitic xenotime → monazite transformation.*

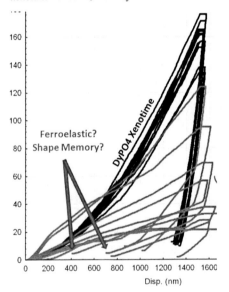

Fig. 9. *Indentation – displacement curves for DyPO₄ (xenotime), and (Gd₀.₅,Dy₀.₅)PO₄ (xenotime). Some (Gd₀.₅,Dy₀.₅)PO₄ indentations show very large displacement rebounds. These are arrowed as ferroelastic or shape memory transformations.*

figure 10. In this figure a TEM micrograph montage is shown along with the relationship of the imaged area to the indenter tip. An area several microns in extent directly beneath the indenter tip has transformed to monazite, as shown by selected area electron diffraction patterns from an area outside (1) and inside (2) the transformed area. This area was imaged exactly one week after the indentation experiments were performed. Subsequent TEM examination months after the indentation experiments found that some of the monazite region had transformed back to xenotime. Time dependent transformation reversal has been observed for other pressure induced martensitic phase transformations under indentation such as β-eucryptite, and in some cases the effect has been correlated with environmental effects such as subcritical crack growth from moisture.[55] The lack of a transformed region in some FIB sections may then be due to either time-dependent reversal of a transformed area between the time of indentation and TEM examination, or lack of any original transformed area. The high variance in load-displacement curves (Fig, 8 & 9) diagnostic of transformation plasticity anisotropy suggests transformation plasticity may not occur under all indentations. We did not track particular load-displacement curves with the indentation FIB sections, so we cannot confirm the anisotropy effect. Work is in progress to identify the transformation reversal mechanisms and time dependence.

Single crystal alumina fibers were coated with TbPO₄ and with (Gd₀.₅,Dy₀.₅)PO₄ for fiber pushout experiments. Temperatures of at least 1400°C were required to deposit these compositions as the xenotime phase in thin film form; this was at least 200°C higher than required for the powders. As observed with the sintered pellets (Fig. 7),

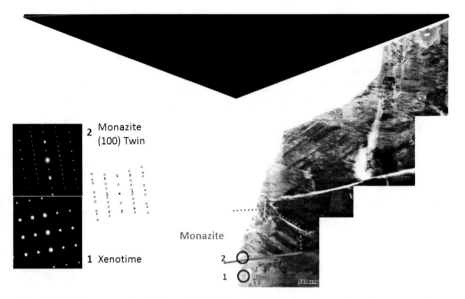

Fig. 10. *TEM micrograph of indented TbPO$_4$ (xenotime), showing transformation to monazite underneath the indent with selected area electron diffraction patterns of the circled areas (1) & (2). The TEM cross-section was prepared by FIB.*

there were significant densification problems that appear to be associated with exaggerated grain growth and loss of phosphorous. The porous coatings were easily infiltrated by the Sumitomo alumina powder and were not suitable for high quality fiber pushout specimens. These processing difficulties are currently under study; we will report more thorough description, characterization, and analysis in a future publication.

If the processing obstacles to making dense coatings of transformable rare-earth orthophosphates are solved, the potential to tailor the fiber pullout friction in CMCs suggested by the large variation in indentation load-displacement for rare-earth orthophosphates and their solid-solutions is particularly attractive for CMC fiber-matrix interphase engineering. Reversal of the transformation with time may also be a desirable feature; the fiber-matrix interphase material resets for applications where load is intermittently applied.

SUMMARY AND CONCLUSIONS

Transformation plasticity of TbPO$_4$ and (Gd$_{0.5}$,Dy$_{0.5}$)PO$_4$ rare-earth orthophosphates with the xenotime phase was demonstrated by nano-indentation and TEM characterization. The martensitic xenotime $\rightarrow$ monazite phase transformations are mechanistically similar to deformation twinning in monazite and require only very small shuffles. There is some evidence that these transformations may reverse with time; this complicates TEM characterization of transformation extent and mechanisms. The TbPO$_4$ and (Gd$_{0.5}$,Dy$_{0.5}$)PO$_4$ compositions are difficult to process as dense pellets or coatings; this currently hinders their use as fiber-matrix interphases in CMCs. Work is in progress to solve the processing difficulties, and to thoroughly characterize the transformation anisotropy and reversal mechanisms.

REFERENCES
[1] Morgan, P. E. D. and Marshall, D. B., Ceramic Composites of Monazite and Alumina, *J. Ceram. Soc.* 78 (6), 1553-63, 1995.
[2] Morgan, P. E. D., Marshall, D. B., and Housley, R. M., High Temperature Stability of Monazite-Alumina Composites, *Mater. Sci. Eng.* A195, 215-222, 1995.
[3] Davis, J. B., Marshall, D. B., Oka, K. S., Housley, R. M., and Morgan, P. E. D., Ceramic Composites for Thermal Protection Systems, *Composites* A30, 483-488, 1999.
[4] Marshall, D. B., Morgan, P. E. D., Housley, R. M., and Cheung, J. T., High-Temperature Stability of the Al_2O_3-$LaPO_4$ System, *J. Am. Ceram. Soc.* 81 (4), 951-956, 1998.
[5] Keller, K. A., Mah, T., Parthasarathy, T. A., Boakye, E. E., Mogilevsky, P., and Cinibulk, M. K., Effectiveness of Monazite Coatings in Oxide/Oxide Composites After Long Term Exposure at High Temperature, *J. Am. Ceram. Soc.* 86 (2), 325-332, 2003.
[6] Lee, P.-Y., Imai, M., and Yano, T., Fracture Behavior of Monazite-Coated Alumina Fiber-Reinforced Alumina-Matrix Composites at Elevated Temperature, *J. Ceram. Soc. Japan* 112 (12), 628-633, 2004.
[7] Marshall, D. B. and Davis, J. B., Ceramics for Future Power Generation Technology: Fiber Reinforced Oxide Composites, *Curr. Opin. Solid State Mater. Sci.* 5, 283-289, 2001.
[8] Kaya, C., Butler, E. G., Selcuk, A., Boccaccini, A. R., and Lewis, M. H., Mullite (Nextel™ 720) Fibre-Reinforced Mullite Matrix Composites Exhibiting Favourable Thermomechanical Properties, *J. Eur. Ceram. Soc.* 22, 2333-2342, 2002.
[9] Davis, J. B., Marshall, D. B., and Morgan, P. E. D., Oxide Composites of $LaPO_4$ and Al_2O_3, *J. Eur. Ceram. Soc.* 19, 2421-2426, 1999.
[10] Davis, J. B., Marshall, D. B., and Morgan, P. E. D., Monazite Containing Oxide-Oxide Composites, *J. Eur. Ceram. Soc.* 20 (5), 583-587, 2000.
[11] Zok, F. W., Developments in Oxide Fiber Composites, *J. Am. Ceram. Soc.* 89 (11), 3309-3324, 2006.
[12] Ruggles-Wrenn, M. B. and Laffey, P. D., Creep Behavior of Nextel™ 720/Alumina Ceramic Composite at Elevated Temperature in Air and in Steam, *Compos. Sci. Tech.* 68, 2260-2266, 2008.
[13] Ruggles-Wrenn, M. B., Siegert, G. T., and Back, S. S., Creep Behavior of NextelTM 720/Alumina Ceramic Composite with +/-45 Fiber Orientation at 1200 C, *Compos. Sci. Tech.* 68, 1588-1595, 2008.
[14] Boakye, E., Hay, R. S., and Petry, M. D., Continuous Coating of Oxide Fiber Tows Using Liquid Precursors: Monazite Coatings on Nextel 720, *J. Am. Ceram. Soc.* 82 (9), 2321-2331, 1999.
[15] Boakye, E. E., Hay, R. S., Mogilevsky, P., and Douglas, L. M., Monazite Coatings on Fibers: II, Coating without Strength Degradation, *J. Am. Ceram. Soc.* 84 (12), 2793-2801, 2001.
[16] Hay, R. S., Boakye, E. E., and Petry, M. D., Effect of Coating Deposition Temperature on Monazite Coated Fiber, *J. Eur. Ceram. Soc.* 20, 589-97, 2000.
[17] Hay, R. S. and Boakye, E., Monazite Coatings on Fibers: I, Effect of Temperature and Alumina Doping on Coated Fiber Tensile Strength, *J. Am. Ceram. Soc.* 84 (12), 2783-2792, 2001.
[18] Kuo, D.-H., Kriven, W. M., and Mackin, T. J., Control of Interfacial Properties through Fiber Coatings: Monazite Coatings in Oxide-Oxide Composites, *J. Am. Ceram. Soc.* 80 (12), 2987-2996, 1997.
[19] Chawla, K. K., Liu, H., Janczak-Rusch, J., and Sambasivan, S., Microstructure and Properties of Monazite ($LaPO_4$) Coated Saphikon Fiber/Alumina Matrix Composites, *J. Eur. Ceram. Soc.* 20, 551-559, 2000.
[20] Cao, H. C., Bischoff, E., Sbaizero, O., Ruhle, M., Evans, A. G., Marshall, D. B., and Brennan, J. J., Effect of Interfaces on the Properties of Fiber-Reinforced Ceramics, *J. Am. Ceram. Soc.* 73 (6), 1691-99, 1990.
[21] Curtin, W. A., Eldredge, J. I., and Srinivasan, G. V., Push-Out Test on a New Silicon Carbide/Reaction Bonded Silicon Carbide Ceramic Matrix Composite, *J. Am. Ceram. Soc.* 76 (9), 2300-2304, 1993.
[22] Kerans, R. J., Hay, R. S., Parthasarathy, T. A., and Cinibulk, M. K., Interface Design for Oxidation Resistant Ceramic Composites, *J. Am. Ceram. Soc.* 85 (11), 2599-2632, 2002.

[23.] Curtin, W. A., Ahn, B. K., and Takeda, N., Modeling Brittle and Tough Stress-Strain Behavior in Unidirectional Ceramic Matrix Composites, *Acta mater.* 46 (10), 3409-3420, 1998.

[24.] Davis, J. B., Hay, R. S., Marshall, D. B., Morgan, P. E. D., and Sayir, A., The Influence of Interfacial Roughness on Fiber Sliding in Oxide Composites with La-Monazite Interphases, *J. Am. Ceram. Soc.* 86 (2), 305-316, 2003.

[25.] Morgan, P. E. D. and Marshall, D. B., Functional Interfaces for Oxide/Oxide Composites, *Mater. Sci. Eng.* A162, 15-25, 1993.

[26.] Sauveur, A., What is Steel? Another Answer, *The Iron Age* 113, 581-583, 1924.

[27.] Wassermann, G., Untersuchungen an einer Eisen-Nickel Legierung uber die Verformbarkeit Wahrend der g-a Unwandlung, *Archiv Fur der Eisenhutt* 7, 321-325, 1937.

[28.] Poirier, J.-P., *Creep of Crystals* Cambridge University Press, Cambridge, 1985.

[29.] Stringfellow, R. G., Parks, D. M., and Olson, G. B., A Constitutive Model for Transformation Plasticity Accompanying Strain-Induced Martensitic Transformations in Metastable Austenitic Steels, *Acta metall. mater.* 40 (7), 1703-1716, 1992.

[30.] Fischer, F. D., A Micromechanical Model for Transformation Plasticity in Steels, *Acta metall. mater.* 38 (6), 1535-1546, 1990.

[31.] Kriven, W. M. and Lee, S.-J., U.S.A 6,361,888, 2002, Toughening of Ceramic Composites by Transformation Weakening of Interphases.

[32.] Kriven, W. M. and Lee, S.-J., Toughening of Mullite/Cordierite Laminated Composites by Transformation Weakening of b-Cristobalite Interphases, *J. Am. Ceram. Soc.* 88 (6), 1521-1528, 2005.

[33.] Poirier, J. P., On Transformation Plasticity, *J. Geophys. Res.* 87 (B8), 6791-6798, 1982.

[34.] Kolitsch, U. and Holtsam, D., Crystal Chemistry of REEXO$_4$ Compounds (X = P, As, V). II. Review of REEXO$_4$ Compounds and their Stability Fields, *Eur. J. Mineral.* 16, 117-126, 2004.

[35.] Jia, C.-J., Sun, L.-D., You, L.-P., Jiang, X.-C., Luo, F., Pang, Y.-C., and Yan, C.-H., Selective Synthesis of Monazite- and Zircon-type LaVO$_4$ Nanocrystals, *J. Phys. Chem. B* 109, 3284-3290, 2005.

[36.] Hay, R. S. and Marshall, D. B., Deformation Twinning in Monazite, *Acta mater.* 51 (18), 5235-5254, 2003.

[37.] Wenk, H.-R., Plasticity Modeling in Minerals and Rocks, in *Texture and Anisotropy*, Kocks, U. F., Tome, C. N., andWenk, H.-R. Cambridge University Press, Cambridge, UK, 1998, pp. 561-596.

[38.] Ushakov, S. V., Helean, K. B., Navrotsky, A., and Boatner, L. A., Thermochemistry of Rare-Earth Orthophosphates, *J. Mater. Res.* 16 (9), 2623-2633, 2001.

[39.] Thiriet, C., Konings, R. J. M., Javorsky, P., and Wastin, F., The Heat Capacity of Cerium Orthophosphate CePO$_4$, the Synthetic Analogue of Monazite, *Phys. Chem. Minerals* 31, 347-352, 2004.

[40.] Popa, K., Sedmidubsky, D., Benes, O., Thiriet, C., and Konings, R. J. M., The High Temperature Heat Capacity of LnPO$_4$ (Ln = La, Ce, Gd) by Drop Calorimetry, *J. Chem. Thermo.*, 2005.

[41.] Thiriet, C., Konings, R. J. M., Javorsky, P., Magnani, N., and Wastin, F., The Low Temperature Heat Capacity of LaPO$_4$ and GdPO$_4$, the Thermodynamic Functions of the Monazite-Type LnPO$_4$ Series, *J. Chem. Thermo.* 37, 131-139, 2005.

[42.] Dorogova, M., Navrotsky, A., and Boatner, L. A., Enthalpies of Formation of Rare Earth Orthovanadates, *REVO$_4$*, *J. Solid State Chem.* 180, 847-851, 2007.

[43.] Mogilevsky, P., Boakye, E. E., and Hay, R. S., Solid Solubility and Thermal Expansion in LaPO$_4$-YPO$_4$ System, *J. Am. Ceram. Soc.* 90 (6), 1899-1907, 2007.

[44.] Keller, K. A., Mogilevsky, P., Parthasarathy, T. A., Lee, H. D., and Mah, T.-I., Monazite coatings in dense ($\geq$90%) alumina-chromia minicomposites, *J. Am. Ceram. Soc.* submitted, 2007.

[45.] Bradby, J. E., Williams, J. S., Wong-Leung, J., Swain, M. V., and Munroe, P., Mechanical Deformation in Silicon by Micro-Indentation, *J. Mater. Res.* 16 (5), 1500-1507, 2001.

[46.] Domnich, V., Gogotsi, Y., and Dub, S., Effect of Phase Transformations on the Shape of the Unloading Curve in the Nanoindentation of Silicon, *Appl. Phys. Lett.* 76 (16), 2214-2217, 2000.

[47.] Zarudi, I., Zhang, L. C., and Swain, M. V., Behavior of Monocrystalline Silicon Under Cyclic Microindentations with a Spherical Indenter, *Appl. Phys. Lett.* 82 (7), 1027-1029, 2003.

[48.] Zhang, L. and Zarudi, I., Towards a Deeper Understanding of Plastic Deformation in Mono-crystalline Silicon, *Int. J. Mech. Sci.* 43, 1985-1996, 2001.

[49.] Boakye, E. E., Hay, R. S., Mogilevsky, P., and Cinibulk, M. K., Two Phase Monazite/Xenotime 30LaPO$_4$-70YPO$_4$ Coating of Ceramic Fiber Tows, *J. Am. Ceram. Soc* 91 (1), 17-25, 2008.

[50.] Boakye, E. E., Fair, G. E., Mogilevsky, P., and Hay, R. S., Synthesis and Phase Composition of Lanthanide Phosphate Nanoparticles LnPO$_4$ (Ln = La, Gd, Tb, Dy, Y) and Solid Solutions for Fiber Coatings, *J. Am. Ceram. Soc.* 91 (12), 3841-3849, 2008.

[51.] Fair, G. E., Hay, R. S., and Boakye, E. E., Precipitation Coating of Monazite on Woven Ceramic Fibers – I. Feasibility, *J. Am. Ceram. Soc.* 90 (2), 448-455, 2007.

[52.] Fair, G. E., Hay, R. S., and Boakye, E. E., Precipitation Coating of Rare-Earth Orthophosphates on Woven Ceramic Fibers- Effect of Rare-Earth Cation on Coating Morphology and Coated Fiber Strength, *J. Am. Ceram. Soc* 91 (7), 2117-2123, 2008.

[53.] Fair, G. E., Hay, R. S., and Boakye, E. E., Precipitation Coating of Monazite on Woven Ceramic Fibers – II. Effect of Processing Conditions on Coating Morphology and Strength Retention of Nextel™ 610 and 720 Fibers, *J. Am. Ceram. Soc.* 91 (5), 1508-1516, 2008.

[54.] Salje, E. K. H., *Phase Transitions in Ferroelastic and Co-Elastic Crystals* Cambridge University Press, 1990.

[55.] Reimanis, I. E., Seick, C., Fitzpatrick, K., Fuller, E. R., and Landin, S., Spontaneous Ejecta from b-Eucryptite Composites, *J. Am. Ceram. Soc.* 90 (8), 2497-2501, 2007.

SOLVOTHERMAL SYNTHESIS OF GADOLINIUM HYDROXIDE AND OXIDE POWDERS AND THEIR POTENTIAL FOR BIOMEDICAL APPLICATIONS

Eva Hemmer
Tokyo University of Science, Department of Materials Science and Technology
2641 Yamazaki, 278-8510 Chiba, Japan

Yvonne Kohl
Fraunhofer Institute for Biomedical Engineering, Department of Biohybrid Systems
Ensheimer Str. 48, 66386 St. Ingbert, Germany

Sanjay Mathur
Institute of Inorganic Chemistry, University of Cologne
Greinstr. 6, 50939 Cologne, Germany

Hagen Thielecke
Fraunhofer Institute for Biomedical Engineering, Department of Biohybrid Systems
Ensheimer Str. 48, 66386 St. Ingbert, Germany

Kohei Soga
Tokyo University of Science, Department of Materials Science and Technology
2641 Yamazaki, 278-8510 Chiba, Japan

ABSTRACT
    The aim of this study was to synthesize gadolinium hydroxide and oxide nanostructures and to assess the toxicity of these materials with regard to potential applications as biomarkers. Gadolinium hydroxide nanorods with several 100 nm length and approximately 30 nm diameter were prepared by solvothermal decomposition of gadolinium oleate precursor. In a second step these structures were doped with europium ions followed by a post-thermal treatment which resulted in crystalline Eu^{3+}:Gd_2O_3 nanostructures. The obtained structures were analyzed with regard to their elemental composition, phase and crystallinity (FT-IR spectroscopy, XRD), size and morphology (SEM, TEM) as well as optical properties (photoluminescence spectroscopy). We investigated the cytotoxic effects of undoped and Eu^{3+}-doped $Gd(OH)_3$ and Gd_2O_3 nanostructures on human colon adenocarcinoma cells (Caco2). Different nanostructure concentrations were tested using cytotoxicity assays (WST-1, LDH) and live/dead staining. As observed in both assays, undoped as well as Eu^{3+}-doped powders do not induce a significant cytotoxic effect in Vitro. These results indicate that the synthesized gadolinium containing inorganic nanostructures are promising candidates for further investigations whose challenge is the development of a new class of contrast agents which may overcome the disadvantages of recently used compounds.

INTRODUCTION
    In recent years, nanomaterials have emerged as important players in modern medicine with applications ranging from contrast agents in medical imaging to carriers for gene delivery into individual cells. Nanoparticles have a number of properties that distinguish them from bulk materials simply by virtue of their size, such as chemical reactivity, energy absorption and biological mobility. Magnetic resonance imaging (MRI) is a technique used to perform 3-D, non-invasive scans of the body. This technique has been recognized as one of the important techniques in medical diagnosis. Nanoparticles can provide significant improvements in MRI of various parts of the body. For MRI an accurate diagnosis is obtained by using positive contrast agents such as ionic gadolinium based

materials in order to enhance the contrast of images in living tissue. Clinically available products are mostly low molecular weight Gd(III) chelates, including DTPA, Gd-DOTA and their derivates.

The responses of cells, cellular components and tissues towards manufactured nano-sized particles as well as their transfer across phase boundaries represented by cellular biological barriers exhibit an important research domain. The application possibilities of nanostructured materials in medicine have been widely explored due their intrinsic properties [1, 2], e.g. their ability to transfer biological barriers (brain-blood-barrier) which qualify them as candidates for diagnosis and drug delivery systems [3]. With regard to new imaging diagnostic tools, materials showing outstanding magnetic and luminescent properties are of growing interest. In this context, lanthanide containing, inorganic nanomaterials are promising candidates for applications as bio-sensors or markers due to their high emission intensity and photostability, whereas the organic dyes used in current fluorescence bioimaging suffer from colour fading resulting in temporal limited use [4]. Semiconductor nanoparticles (quantum dots) are reported to be a promising alternative to recently used organic dyes [5-7]. Their main disadvantage is their toxicity [8] which requires a coating resulting in core-shell structures like CdSe/ZnS [9]. Consequently, there is a demand for the development of new biocompatible materials. The use of lanthanide containing inorganic materials is expected to make a significant impact as advanced contrast agents (CAs) for MRI. Gadolinium oxide is a promising alternative to recently used contrast agents in computer tomography and MRI, due to its excellent X-ray absorption properties, high content of radiopaque metal [10] and selective liver-spleen clearance and paramagnetic properties [11]. Despite of an increasing interest in multifunctional nanomaterials for biomedical applications [12, 13], there is still a lack of understanding of the interactions between manufactured nanomaterials and biological tissue. However, for an effective and safe use of nanostructures in biomedical applications risk assessment and management is required [14-16] and numerous studies addressed the toxicology of nanostructures in recent years [17-19]. For metal oxides, mainly cytotoxicity of TiO_2, ZrO_2 or Al_2O_3 nanoparticles has been investigated [20-22]. With regard to lanthanide containing compounds, gadolinium-based MRI contrast agents [23, 24], $GdCl_3$ [25] or gadolinium-based redox active drug for cancer therapy [26] have been studied. Heinrich et al. analyzed the cytotoxicity of iodinated and gadolinium-based contrast agents [27]. He found out, that gadolinium-based contrast agents do not induce fewer cytotoxic effects on renal proximal tubular cells than do iodinated and no significant difference in induced necrosis or apoptosis between the two contrast agents had been detected. However, only few reports about Ln_2O_3 or $Ln(OH)_3$ have been published [28, 29]. Our interest lies in the preparation of lanthanide containing nanocrystalline structures with respect to future applications in biomedicine. In this study, we focus on undoped and europium-doped gadolinium hydroxide ($Gd(OH)_3$ and Eu^{3+}:$Gd(OH)_3$) and oxide structures (Eu^{3+}:Gd_2O_3). In a first step, the obtained powders were analysed with regard to composition, phase and morphology. In order to investigate the optical properties of the doped samples photoluminescence excitation and emission spectra were recorded. Due to our special interest in development of lanthanide containing inorganic materials for biomedical applications the second main aspect is the investigation of their cytotoxic behavior in vitro.

EXPERIMENTAL DETAILS

Material Synthesis and Characterization

As reported in literature, metal oleate complexes are suitable starting compounds for the synthesis of nanoscale metals, metal oxides or metal sulfides [30-32]. Gadolinium hydroxide ($Gd(OH_3)$) nanorods were obtained by solvothermal decomposition of gadolinium oleate [33]. The precursor gadolinium oleate was synthesized by salt elimination reaction of gadolinium chloride hexahydrate (2.5 mg, 6.8 mmol) with sodium oleate (6.2 mg, 20.3 mmol) following the procedure described by Hyeon et al. [30]. After washing with an ethanol-water mixture (ethanol : water = 1 : 1) the air and moisture stable product was obtained as colourless waxy solid. The good agreement of the results of elemental analysis with calculated carbon and hydrogen content (C: 62.76 % / calculated: 64.75 %, H: 10.12 % / calculated: 9.96 %) and the sharp peaks in the recorded FT-IR spectrum at 1650 to 1510 cm^{-1} which are characteristic for metal-carboxylate-vibrations confirm the formation of the desired compound.

For material synthesis 200 mg gadolinium oleate complex was dissolved in 20 mL hexane followed by pre-hydrolysis with aqueous potassium hydroxide solution (V = 1 mL, c = 1.87 mol/L). For solvothermal treatment Teflon liners (V = 50 mL) with the activated solution were closed in stainless steel autoclaves (DAB-2, Berghof Products + Instruments GmbH), temperature was raised to 250 °C and kept for 24 hrs. After cooling to room temperature obtained precipitates were washed several times with methanol and ethanol, collected by centrifugation and dried under ambient conditions. The white powders were pound before further characterization. For doping of Gd(OH)$_3$ with Eu^{3+} ions, 250 mg (1.2 mmol) of as-prepared gadolinium hydroxide was dispersed in 10 mL water, followed by the addition of 10 mL of an aqueous solution of EuCl$_3 \cdot 6$ H$_2$O (43.98 mg, 0.12 mmol, doping rate: 10 mol%). 1,2-ethanediol (13.38 mL, 0.24 mmol) as complexing agent was added to the above solution [34]. The solution was concentrated by slow evaporation at 60 °C with stirring resulting in a white gel, which was grounded and annealed in air at 600 and 1100 °C during 2 hrs.

The crystalline phases of the as-prepared and calcinated samples were determined with a powder X-ray diffractometer (ULTIMA III, Rigaku) using CuK$_{\alpha1}$ radiation as the X-ray source. Morphologies were observed by SEM (S-4200, Hitachi) and TEM (JEOL 200 CX, Philips). IR-absorption bands of the precursor were determined using a Vektor 22 spectrometer by Bruker. For FT-IR spectroscopy of the obtained materials powders were mixed with KBr and transmission was recorded under vacuum conditions using a Jasco FT/IR-6200 spectrometer. Emission and excitation spectra of europium-doped samples were recorded at room temperature using a RF-5000 spectrofluorophotometer by Shimadzu-Seisakusho.

Cytotoxicity Tests

Cytotoxicity of the obtained nanostructures was evaluated using a consistent set of in vitro experimental protocols. Human colon adenocarcinoma cells (Caco2, American Type Culture Collection, Manassas, VA) were used for the in vitro study of the nanostructure toxicology. Caco2 cells were cultured in DMEM/F-12 (1 : 1) (Invitrogen, Carlsbad, California), supplemented with 20 % FCS, 100 U/mL penicillin and 100 µg/mL streptomycin. All cell culture systems were cultured at 37 °C, 5 % CO$_2$ and 95 % saturated atmospheric humidity.

The mitochondrial function of the incubated cells was analyzed using the WST-1 assay (Roche Diagnostics, Mannheim, Germany). This assay is based on the cleavage of stable tetrazolium salt WST-1 (4-[3-(4-iodophenyl)-2-(4-nitrophenyl)-2H-5-tetrazolio]-1,3-benzol-disulfonate) by metabolically active cells to an orange formazan dye. WST-1 assay was performed after 4 and 24 hrs according to manufacturer's instructions with appropriate controls. As positive control TritonX-100 was used. This detergent is a non-ionic surfactant which has a hydrophilic polyethylene oxide group and a hydrocarbon lipophilic group. Because of its lipophilicity it permeabilizes cell membranes. The incubation of TritonX-100 induces membrane damage and cell death. The incubation of cell culture medium without nanorods demonstrates the negative control in order to simulate the cell behaviour under ideal conditions. The data obtained after the incubation of the nanorods is compared to the negative control. The mentioned controls were utilized in all assays performed in our study.

After exposure to the nanostructures, cells were incubated with the ready-to-use WST-1 reagent for 4 or 24 hrs. After the respective incubation period, the absorbance was quantified using a scanning multi-well spectrophotometer reader (ELISA, Tecan Germany, Crailsheim). Lactate dehydrogenase (LDH) leakage, which is another measure of cytotoxicity on the basis of membrane integrity damage, was determined using a LDH kit (Roche Diagnostics, Mannheim, Germany). Released LDH catalyzed the oxidation of lactate to pyruvate with simultaneous reduction of NAD^+ to NADH. The rate of NAD^+ reduction was measured as an increase in absorbance at 340 nm. The rate of NAD^+ reduction was directly proportional to LDH activity in the cell medium. After incubation with nanostructures for 4 and 24 hrs, the cell culture medium was collected for LDH measurement. In order to detect the percentage of viable and dead cell, a dye mixture of fluoresceine diacetate (FD) and propidium iodide (PI) was employed to stain the cells. The cells were grown in a cell culture dish (35/10 mm, Greiner Bio One, Frickenhausen, Germany). Cell culture medium containing nanomaterial was incubated for 4 and 24 hrs. Afterwards, the cells were treated with the fluorescent dye mixture and observed via confocal fluorescence microscope (Zeiss). Living cells show green fluorescence, while dead cells show a red fluorescence signal.

RESULTS AND DISCUSSION

Material Synthesis and Characterisation
General advantages of the solvothermal synthesis when compared to conventional preparation of ceramic materials like sol-gel techniques are the low reaction temperature (< 300 °C), nanoscalic particles with homogeneous size distribution and high phase purity [35, 36]. TEM micrograph from powders obtained by solvothermal decomposition of gadolinium oleate precursor is shown in figure 1a.

(a)   (b)
Figure 1. TEM image of (a) as-prepared Gd(OH)$_3$ nanorods obtained by solvothermal treatment of Gd(oleate)$_3$ and (b) FE-SEM image of Eu^{3+}-doped Gd$_2$O$_3$ sample (annealed at 600 °C).

Elongated structures with homogeneous length of several hundred nanometers and diameter of approximately 30 nm are observed. Due to the ratio of length and diameter (aspect ratio: 15 - 20) the obtained nanostructures can be classified as nanorods. The lattice fringes are observed in high resolution TEM and confirm the good crystallinity of the material. As shown by XRD measurements the obtained nanorods consist of crystalline gadolinium hydroxide, Gd(OH)$_3$ (figure 2a). Gadolinium hydroxide crystallizes in the hexagonal lattice (a = 6.3290 Å, b = 6.3290 Å, c = 3.6310 Å). Therefore, anisotropic growth is favoured by the intrinsic material properties and consequently no templates are required to prepare elongated structures. Doping of the as-prepared Gd(OH)$_3$ nanorods with 10 mol%

Eu^{3+} ions does not take influence on phase or crystallinity. The X-ray powder diffractogram recorded on the doped material shows the characteristic reflexes for crystalline $Gd(OH)_3$ (figure 2b). Post thermal treatment of Eu^{3+}:$Gd(OH)_3$ at 600 °C for 2 hrs led to oxidation of the hydroxide phase to gadolinium oxide, Gd_2O_3, as shown by X-ray powder diffraction. Investigation of the obtained oxide powder in FE-SEM reveals an inhomogeneous morphology. Besides nanorods of more than 100 nm in length and approximately 50 nm in diameter, agglomerations of irregular shaped particles are observed (figure 1b).

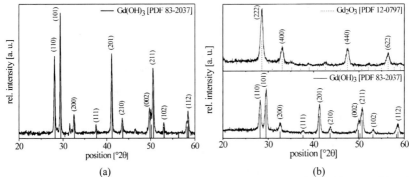

(a)   (b)

Figure 2. X-ray diffraction patterns of (a) the sample obtained by solvothermal decomposition of gadolinium oleate and (b) as-prepared and at 600 °C annealed europium doped powders.

Elimination of hydroxyl groups with increasing annealing temperature can be observed by FT-IR spectroscopy recorded for as-prepared and post-annealed samples (figure 3). The FT-IR spectrum obtained for the as-prepared Eu^{3+}:$Gd(OH)_3$ shows a strong peak and band in the range from 3600 to 3100 cm^{-1} which can be related to vibration of hydroxyl groups. Further peaks in the range from 1650 to 1510 cm^{-1} can be assigned to metal carboxylate vibration which may be due to residual oleate groups originating from $Gd(oleat)_3$ precursor.

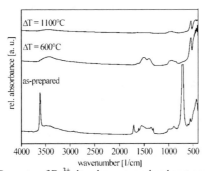

Figure 3. FT-IR spectra of Eu^{3+}-doped as-prepared and post-annealed samples.

Heat treatment of the sample removes the water and carboxylate groups from the powder and no bands due to water or carboxylate groups are observed after calcination at 1100 °C. For calcinated powders a peak at 540 cm^{-1} is observed which can be assigned to M-O-vibration indicating the formation of metal oxide phase.

Excitation and emission spectra of the europium-doped hydroxide and oxide powders are given in figure 4. All spectra show the characteristic peaks corresponding to Eu^{3+} f-f transitions. The excitation spectrum of Eu^{3+}:Gd_2O_3 powders annealed at 600 °C shows three intensive absorption peaks at 393.3, 465.2 and 531.6 nm at a monitoring wavelength of 611.7 nm (figure 4a). The observed bands are referred to $^7F_0 \rightarrow ^5L_6$ (393.3 nm), $^7F_0 \rightarrow ^5D_2$ (465.2 nm) and $^7F_{0,1} \rightarrow ^5D_1$ (393.3 nm) transitions. Emission spectra were recorded under an excitation at 393.3 nm. Figure 4b shows the emission spectra recorded for Eu^{3+}-doped gadolinium hydroxide and oxide in dependence of annealing temperature. In all cases a maximum peak in the red range of the visible spectrum at 610 nm is observed which is due to the $^5D_0 \rightarrow ^7F_2$ transition. Further peaks are observed at 578, 590, 626, 648 and 704 nm which are assigned to $^5D_0 \rightarrow ^7F_J$ (J = 0, 1, 2, 3, 4) transitions. The ratio between the intensity of electric dipole emission (I_{ED}) and the intensity of magnetic dipole emission (I_{MD}), $I_{ED} / I_{MD} = I(^5D_0 \rightarrow ^7F_2) / I(^5D_0 \rightarrow ^7F_1)$, increases with increasing annealing temperature. The I_{ED} is strongly affected by the surrounding of the Eu^{3+} ion to be increased by increasing the covalency with the charge of the ligands to give stronger crystal field, while the I_{MD} is almost unaffected by the change of matrices. The observation of the emission spectra can be attributed to the change from monovalent OH^- to divalent O^{2-} in the environment of luminescent Eu^{3+} centres when hydroxide is transformed into oxide resulting in changes in the crystal field. Herein, powders obtained at 600 °C represent an intermediate state between the pure hydroxide (as-prepared sample) and oxide (sample annealed at 1100 °C) phase.

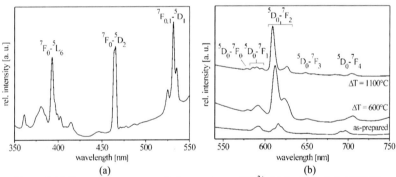

Figure 4. (a) Room temperature excitation spectrum of Eu^{3+}:Gd_2O_3 (annealed at 600 °C, λ_{em} = 611.7 nm) and (b) emission spectra of Eu^{3+}:$Gd(OH)_3$ as well as Eu^{3+}:Gd_2O_3 (λ_{ex} = 393.3 nm).

Cytotoxicity Tests

With regard to potential applications in biomedicine the effect of undoped and europium-doped gadolinium hydroxide ($Gd(OH)_3$ and Eu^{3+}:$Gd(OH)_3$) and oxide nanostructures (Eu^{3+}:Gd_2O_3) in human colon adenocarcinoma cells was determined. The nanomaterial-induced cytotoxic effects were evaluated quantitatively utilizing lactate dehydrogenase (LDH) and water soluble formazane (WST-1) assays.

The mitochondrial function of the Caco2 cells was measured via the WST-1 assay after culturing in presence of undoped and doped gadolinium-based nanorods for two different incubation times (4 and 24 hrs). In case of all three samples, no significant decrease of the mitochondrial activity was observed at a nanorod concentration of 50 µg/mL (figure 5a). As also evident from figure 5a, at a concentration of 50 µg/mL, no time-dependent effect was detected. Even a higher concentration of 250 µg/mL Gd(OH)$_3$, Eu^{3+}:Gd(OH)$_3$) and Eu^{3+}:Gd$_2$O$_3$ induced no significant decrease of the mitochondrial activity, when compared to the negative control (figure 5b). Compared to the effect after the incubation at a nanorod concentration of 50 µg/mL a light decrease in mitochondrial function was observed for the higher concentration. However, this decrease of less than 20 % is not significant. No significant difference between the two incubation times was determined.

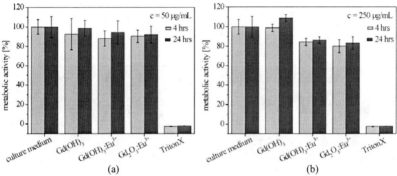

(a)   (b)

Figure 5. Metabolic activity of Caco2 cells incubated with (a) 50 µg/mL and (b) 250 µg/mL gadolinium containing nanostructures.

The membrane integrity of the Caco2 cells incubated with europium-doped nanostructures was measured by means of the LDH assay. Following the exposure of our samples at highest dosage level, LDH release was not significantly higher than control (figure 6). Further, no time-dependent effect on the plasma membrane at any of the tested concentrations could be observed.

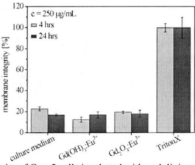

Figure 6. Membrane integrity of Caco2 cells incubated with gadolinium containing nanostructures (c = 250 µg/mL).

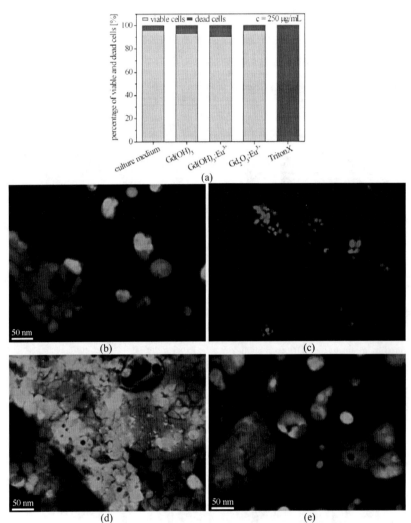

(a)

(b)   (c)

(d)   (e)

Figure 7. Analysis of the percentage of viable and dead Caco2 cells incubated with gadolinium containing nanostructures (a) and fluorescence images of Caco2 cells after fluoresceine diacetat staining incubated with (b) pure medium, (c) 2 % TritonX, (d) Gd(OH)$_3$ and (e) Eu^{3+}:Gd$_2$O$_3$ (c = 250 µg/mL).

Additionally to the cytotoxicity tests a live/dead staining was performed. All fluorescence images were taken after 24 hrs of exposure. The observed results correlate with the results obtained from cytotoxicity assays (figure 7a). The incubation of cell culture medium without nanorods shows a low percentage of dead Caco2 cells. 95.69 % of Caco2 control cells are viable. After the exposure to 250 µg/mL Gd(OH)$_3$ the amount of dead cells increased by 2.60 % to 6.91 % compared to the control. Addition of 250 µg/mL of Eu^{3+}:Gd(OH)$_3$ induced an increase of dead cells to 9.71 %. In case of Eu^{3+}:Gd$_2$O$_3$ an increase to 4.38 % was observed. Fluorescence microscopy images of Caco2 cells after 24 hrs incubation with Gd(OH)$_3$ and Eu^{3+}:Gd$_2$O$_3$ (c = 250 µg/mL) are shown in figure 7d and e. When compared to Caco2 control cells (figure 7b and c) the high percentage of viable cells is confirmed.

According to the above described results of WST-1 and LDH assay as well as live/dead staining, the presented gadolinium hydroxide and oxide nanostructure do not induce any cytotoxic effect, independent from incubation time till the highest tested concentration.

CONCLUSION

Undoped and Eu^{3+}-doped gadolinium hydroxide and oxide nanostructures (Gd(OH)$_3$, Eu^{3+}:Gd(OH)$_3$ and Eu^{3+}:Gd$_2$O$_3$) have been synthesized using gadolinium oleate as precursor. Its solvothermal decomposition led to crystalline Gd(OH)$_3$ with a homogeneous nanorod-like morphology of several 100 nm in length and approximately 30 nm in diameter. After doping with Eu^{3+} ions and post-thermal treatment of the as-prepared gadolinium hydroxide nanorods crystalline Eu^{3+}:Gd$_2$O$_3$ nanostructures were obtained. The rod-like morphology could partly be prevented, but additional agglomerations of inhomogeneous structures were observed in FE-SEM. The obtained europium containing structures were investigated with regard to their optical properties recording photoluminescence spectra. Excitation and emission spectra showed peaks characteristic for the f-f transitions of Eu^{3+} ions.

Cytotoxicity tests were affected in order to investigate potential inhibitory effects of the prepared nanostructures on cell functions. WST-1 assay didn't show any effect on the metabolic activity of the Caco2 cells. The greatest viability loss (19.86 %) was found at 250 µg/mL after the incubation with Eu^{3+}:Gd$_2$O$_3$ for 4 as well as 24 hrs (16.73 %). None of the investigated samples induced an apparent LDH leakage from Caco2 cells treated for 4 and 24 hrs which reveals no impact of the nanorods on cell membrane integrity. The Caco2 cells were still viable after the incubation with Gd(OH)$_3$, Eu^{3+}:Gd(OH)$_3$ and Eu^{3+}:Gd$_2$O$_3$ as confirmed by live/dead staining. These results make Eu^{3+}-doped gadolinium hydroxide and oxide to a promising alternative to recently used contrast agents and fluorescent markers.

With regard to future applications requiring well dispersed and homogeneous nanostructures, our future work is focused on the optimization of the synthesis route and investigation of suitable surface modifications.

ACKNOWLEDGEMENT
The authors thank the German Science Foundation (DFG) for the financial support in the frame of the Schwerpunktprogramm SPP-1166.

REFERENCES
[1] V. Wagner, A. Dullaart, A.-K. Bock and A. Zweck, The Emerging Nanomedicine Landscape, *Nat. Biotechnol.*, **24** [10], 1211-1217 (2006).
[2] X. Chen and H. J. Schluesener, Nanosilver: A Nanoproduct in Medical Application, *Toxicol. Lett.*, **176** [1], 1-12, (2008).
[3] E. Angeli, R. Buzio, G. Firpo, R. Magrassi, V. Mussi, L. Repetto and U. Valbusa, Nanotechnology Applications in Medicine, *Tumori.*, **94** [2], 206-215 (2008).

[4] N. C. Shaner, M. Z. Lin, M. R. McKeown, P. A. Steinbach, K. L. Hazelwood, M. W. Davidson and R. Y. Tsien, Improving the Photostability of Bright Monomeric Orange and Red Fluorescent Proteins, *Nature Methods*, **5**, 545-551 (2008).

[5] M. Bruchez, Jr., M. Moronne, P. Gin, S. Weiss and A. P. Alivisatos, Semiconductor Nanocrystals as Fluorescent Biological Labels, *Science*, **281**, 2013-2016 (1998).

[6] W. C. W. Chan and S. Nie, Quantum Dot Bioconjugates for Ultrasensitive Nonisotopic Detection, *Science*, **281**, 2016-2018 (1998).

[7] J. Kim, K. S. Kim, G. Jiang, H. Kang, S. Kim, B.-S. Kim, M. H. Park and S. K. Hahn, In Vivo Real-Time Bioimaging of Hyaluronic Acid Derivatives Using Quantum Dots, *Biopolymers*, **89** [12], 1144-1153 (2008).

[8] N. Lewinski, V. Colvin and R. Drezek, Cytotoxicity of Nanoparticles, *SMALL*, **4**, 26-49 (2008).

[9] B. O. Dabbousi, J. Rodriguez-Viejo, F. V. Mikulec, J. R. Heine, H. Mattoussi, R. Ober, K. F. Jensen and M. G. Bawendi, (CdSe)ZnS Core-Shell Quantum Dots: Synthesis and Characterization of a Size Series of Highly Luminescent Nanocrystallites, *J. Phys. Chem. B*, **101**, 9463-9475 (1997).

[10] A. Havon, M. A. Davis, S. E. Seltzer, A. J. Paskins-Hurlburt and S. J. Hessel, Heavy Metal Particulate Contrast Materials for Computed Tomography of the Liver, *J. Computer Assisted Tomography*, **4** [5], 642-648 (1980).

[11] K. R. Burnett, G. L. Wolf, R. Schumacher and E. J. Goldstein, Gadolinium Oxide: A Prototype Agent for Contrast Enhanced Imaging of the Liver and Spleen with Magnetic Resonance, *Magnetic Resonance Imaging*, **3**, 65-71 (1985).

[12] L. Wang, M. B. O'Donoghue and W. Tan, Nanoparticles for Multiplex Diagnostics and Imaging, *Nanomedicine*, **1**, 413-426 (2006).

[13] W. J. M. Mulder, A. W. Griffioen, G. J. Strijkers, D. P. Cormode, K. Nicolay and Z. A. Fayad, Magnetic and Fluorescent Nanoparticles for Multimodality Imaging, *Nanomedicine*, **2**, 307-324 (2007).

[14] D. B. Resnik and S. S. Tinkle, Ethics in Nanomedicine, *Nanomedicine*, **2** [3], 345-350 (2007).

[15] V. L. Colvin, The Potential Environmental Impact of Engineered Nanomaterials, *Nat. Biotechnol.*, **21** [10], 1166-1170 (2003).

[16] R. Clift, Don't Bury your Head in the Nanostuff, *Nanomedicine*, **2** [3], 267-270 (2007).

[17] G. Oberdörster, E. Oberdörster and J. Oberdörster, Nanotoxicology: An Emerging Discipline Evolving from Studies of Ultrafine Particles, *Environ. Health Persp.*, **113**, 823-839 (2005).

[18] M. Geiser, B. Rothen-Rutishauser, N. Kapp, S. Schürch, W. Kreyling, H. Schulz, M. Semmler, V. Im Hof, J. Heyder and P. Gehr, Ultrafine Particles Cross Cellular Membranes by Nonphagocytic Mechanisms in Lungs and in Cultured Cells, *Environ. Health Persp.*, **113**, 1555-1560 (2005).

[19] H. Yang, C. Liu, D. Yang, H. Zhang and Z. Xi, Comparative Study of Cytotoxicity, Oxidative Stress and Genotoxicity Induced by Four Typical Nanomaterials: The Role of Particle Size, Shape and Composition, *J. Appl. Toxicol.*, **29** [1], 69-78 (2009).

[20] O. R. Moss, Insights into the Health Effects of Nanoparticles: Why Numbers Matter, *Int. J. Nanotechnol.*, **5** [1], 3-14 (2008).

[21] C. M. Sayes and D. B. Warheit, An In Vitro Investigation of the Differential Cytotoxic Responses of Human and Rat Lung Epithelial Cell Lines Using TiO_2 Nanoparticles, *Int. J. Nanotechnol.*, **5** [1], 15-29 (2008).

[22] K. Soto, K. M. Garza and L. E. Murr, Cytotoxic Effects of Aggregated Nanomaterials, *Acta Biomaterialia*, **3**, 351-358 (2007).

[23] A. L. Doiron, K. Chu, A. Ali and L. Brannon-Peppas, Preparation and Initial Characterization of Biodegradable Particles Containing Gadolinium-DTPA Contrast Agent for Enhanced MRI, *PNAS*, **105** [45], 17232-17237 (2008).

[24] J. L. Major, G. Parigi, C. Luchinat and T. J. Meade, The Synthesis and In Vitro Testing of a Zinc-Activated MRI Contrast Agent, *PNAS*, **104** [35], 13881-13886 2007.

[25] D. A. Badger, R. K. Kuester, J.-M. Sauer and I. G. Sipes, Gadolinium Chloride Reduces Cytochrome P450: Relevance to Chemical-Induced Hepatotoxicity, *Toxicology*, **121**, 143-153 (1997).

[26] D. Magda and R. A. Miller, Motexafin Gadolinium: A Novel Redox Active Drug for Cancer Therapy, *Seminars in Cancer Biology*, **16**, 466-476 (2006).

[27] M. C. Heinrich, M. K. Kuhlmann, S. Kohlbacher, M. Scheer, A. Grgic, M. B. Heckmann and M. Uder, Cytotoxicity of Iodinated and Gadolinium-based Contrast Agents in Renal Tubular Cells at Angiographic Concentrations: In Vitro Study, *Radiology*, **242** [2], 425-434 (2007).

[28] S. Hirano and K. T. Suzuki, Exposure, Metabolism, and Toxicity of Rare Earths and Related Compounds, *Environ. Health Persp.*, **104** [Supplement 1], 85-95 (1996).

[29] M. A. McDonald, B. S and K. L. Watkin, Small Particulate Gadolinium Oxide and Gadolinium Oxide Albumin Microspheres as Multimodal Contrast and Therapeutic Agents, *Invest. Radiol.*, **38**, 305-310 (2003).

[30] J. Park, K. An, Y. Hiwang, J.-G. Park, H.-J. Noh, J.-Y. Kim, J.-H. Park, N.-M. Hwang and T. Hyeon, Ultra-Large-Scale Syntheses of Monodisperse Nanocrystals, *Nature Materials*, **3**, 891-895 (2004).

[31] L. M. Bronstein, X. Huang, J. Retrum, A. Schmucker, M. Pink, B. D. Stein and B. Dragnea, Influence of Iron Oleate Complex Structure on Iron Oxide Nanoparticle Formation, *Chem. Mater.*, **19**, 3624-3632 (2007).

[32] S. V. Mahajan and J. H. Dickerson, Synthesis of Monodisperse sub-3 nm RE_2O_3 and $Gd_2O_3{:}RE^{3+}$ Nanocrystals, *Nanotchnology*, **18**, 325605 (6 pp) (2007).

[33] S. Mathur, E. Hemmer, Y. Kohl and H. Thielecke, Cytotoxicity of $Gd(OH)_3$ Nanostructures Prepared by Solvothermal Synthesis, edited by A.-V. Mudring, I. Pantenburg, *Conference Proceedings XXI. Tage der Seltenen Erden - Terrae Rarae 2008*, ISBN 978-3-941372-00-9, NWT-Verlag, Bornheim, Germany (2008).

[34] M. Veith, S. Mathur, A. Kareiva, M. Jilavi, M. Zimmer and V. Huch, Low Temperature Synthesis of Nanocrystalline $Y_3Al_5O_{12}$ (YAG) and Ce-doped $Y_3Al_5O_{12}$ via Different Sol-Gel Methods, *J. Mater. Chem.*, **9**, 3069-3079 (1999).

[35] S. Komarneni, Nanophase Materials by Hydrothermal, Microwave-Hydrothermal and Microwave-Solvothermal Methods, *Current Science*, **85**, 1730-1734 (2003).

[36] L. Cushing, V. L. Kolesnichenko and C. J. O'Connor, Recent Advances in the Liquid-Phase Syntheses of Inorganic Nanoparticles, *Chem. Rev.*, **104**, 3839-3946 (2004).

CVD GROWN SEMICONDUCTOR NANOWIRES: SYNTHESIS, PROPERTIES AND CHALLENGES

J. Pan, H. Shen and S. Mathur
Institute of Inorganic and Materials Chemistry, University of Cologne
D-50939, Cologne
Germany

ABSTRACT

Semiconductor nanowires (NWs) exhibit novel electronic and optical properties owing to their unique structural dimensionality and play a critical role in future electronic and optoelectronic devices. Molecule-based chemical vapor deposition (MB-CVD), based on catalyzed Vapor–Liquid–Solid (VLS) growth mechanism, is an efficient way to synthesize one-dimensional (1D) semiconductor nanostructures. Current research is focused on rational synthesis of 1D nanoscale building blocks, demonstration of novel properties, fabrication of functional device, and integration of nanowire elements into complex functional architectures. Here, we review recent advances in synthesis and properties of semiconductor NWs via CVD, using work from our laboratory for illustration.

INTRODUCTION

Semiconductor nanowires have become the focus of intensive research due to their finite size effects and unique physical properties important for the fabrication of nano-devices.[1] It is generally accepted that 1D nanostructures offer a good system to investigate the dependence of electrical and thermal transport behaviors on dimensionality and size reduction. They also would play an important role as both interconnects and functional units in fabricating electronic and optoelectronic devices on nanoscale.

Semiconductor nanowires represent an important and broad class of nanoscaled wire-like structure, which meanwhile can be rationally and predictably synthesized in single crystalline form with controlled chemical composition, diameter, length, and doping level.[2,3] The availability of nanowires has enabled a wide-range of proto-type devices and integration strategies which however need to be pursued in a rational manner. For example, semiconductor NWs have been assembled into field effect transistors (FETs),[4] p-n diodes,[4,5] light emitting diodes (LEDs),[4] bipolar junction transistors,[6] complementary inverters,[6] complex logic gates and even computational circuits that have been used to carry out basic digital calculations.[7] Semiconductor nanowires have been successfully fabricated by laser ablation,[8] vapor transport,[9] solvothermal,[10] Coulomb blockade (or single-electron tunneling, SET),[11] template-assisted electrochemical[12] and chemical vapor deposition methods.[13,14] Despite the fact that a number of methods have been demonstrated to produce such nanowires, the quest for a generic approach still continues. Among several synthetic approaches demonstrated for nanowire synthesis, CVD appears to be the most attractive candidate because of some distinctive advantages, such as:[15] (i) the capability of producing highly dense and pure materials; (ii) reproducibility of the syntheses and better adhesion of materials on substrates; (iii) high growth rates and coating of complex shaped components, and (iv) the ability to control crystal structure, surface morphology and orientation of nanostructures by controlling the process parameters. Molecule-based CVD technique relying on the decomposition of single molecular precursors allows to synthesize different nanostructures under controlling of chemical composition, size, morphology and surface state at relatively low temperature.

This paper will mainly focus on the synthesis, properties and applications of semiconductor nanowires obtained by molecule-based CVD and illustrate the technical issues relevant to the semiconductor nanowires fabrication, identify measurements and insights needed to push the development of nanowire based heterostructures forward, and suggest new types of devices enabled by the developments in 1D synthesis.

SYNTHESIS AND PROPERTIES OF NANOWIRES

Semiconductor NWs are generally synthesized by using metallic catalysts via a vapor-liquid-solid (VLS) growth process (Scheme 1).[16] During this growth process, the metal nanoparticles are heated above the eutectic temperature for the metal-semiconductor system in the presence of a vapor-phase source of the semiconductor, resulting in a liquid droplet of the metal/semiconductor composite. The semiconductor source, which is delivered by the fragment of molecular precursor, feeds the liquid droplet continuously to supersaturate the eutectic through the vapor-liquid (V-L) interface. It leads to nucleation (crystallization) of the solid semiconductor NWs. The liquid-solid (L-S) interface, which forms the growth interface and acts as a sink causing the continued semiconductor incorporation into the lattice and, thereby, the growth of the nanowires with the alloy droplets riding on the top. Continuous vapor delivery provides the driving force for diffusion of the semiconductor from the liquid-catalyst particle surface to the growth interface. In principle, vapor-liquid-solid growth technique is a simple process in which condensed or powder source material is vaporized at elevating temperature and then the resultant vapor phase condenses at certain conditions (temperature, pressure, atmosphere, substrate, etc.) to form the expected 1D product. The chemical and physical properties of sources and substrates can influence the growth behavior of 1D nanostructures via the interaction with V-L and L-S growth fronts (Scheme 1).

Au Particles   Alloy Liquid   Nucleation of NWs   NW Growth   Substrate

Scheme 1. The controlled growth of nanowires by regulation of chemcial and physical parameters.

The gaseous semiconductor reactants are generated through decomposition of precursors in a chemical vapor deposition process. Controlled gas-phase reaction chemistry and decomposition behavior of metal-organic precursors have proved to be of great value in the chemically governed synthesis of pure thin films[17] and is equally applicable to nanowire synthesis.[18,19,20] The choice of molecular precursor is usually based on the stability of the metal-ligand bond, convenient handing, and the possibility to use a single-source to insure a controlled and high feedstock in the gas phase. In most of the cases, metal-ligand connectivity, metal ratios and valences are conserved during the precursor-to-material conversion, which lead to chemically control the process of material formation (Scheme 2). A number of tailored one-dimensional nanostructures have been fabricated by molecule-based CVD, such as nanowires (Ge,[18] SnO_2,[20] Fe_3O_4, V_2O_5, In_2O_3), core-shell structures (Ge/SiC$_x$N$_y$,[19] $SnO_2@TiO_2$, $SnO_2@Fe_3O_4$) and heterostructures ($SnO_2@SnO_2$, $SnO_2@VO_x$).

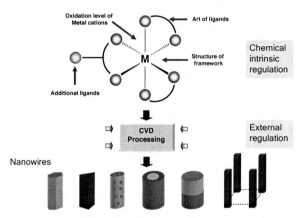

Scheme 2. Processing of 1D nanostructures by molecule-based CVD.

Conventional material synthesis procedures, such as laser ablation and vapor transport reactions, require high-temperatures (> 800°C); on the other hand the molecule-based CVD technique offers low-temperature alternative for synthesis of single crystalline nanowires. Single crystalline Ge NWs were successfully synthesized by CVD of a molecular precursor, $[Ge(C_5H_5)_2]$.[18] The gas phase species were detected by on-line mass spectrometry analysis of the volatile by-products (eq. 1).

$$[Ge(C_5H_5)_2] \rightarrow Ge^+ + C_xH_y^+ + H_2^+ \tag{1}$$

The decomposition of $[Ge(C_5H_5)_2]$ on a Au-coated Si(100) produced already at 300°C the homogeneous deposits of Ge nanowires (Fig. 1a). It is a viable lower temperature route than other conventional approaches. The high-resolution TEM image (Fig. 1b) revealed defect-free single crystalline structure with a preferred growth direction, which was indexed to be [11-2] by FFT analysis (Inset, Fig. 1b).

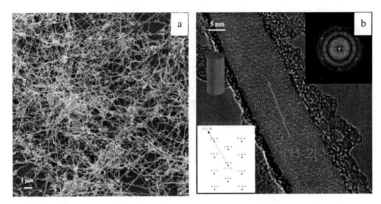

Figure 1. SEM and TEM images of Ge nanowires.

The Ge nanowires obtained by conventional methods usually contain germanium oxide on the surface. Although the pure Ge nanowires can be synthesized by CVD, however, the oxidation of Ge nanowires can not be avoided in air and the solubility of germanium oxide in water. To overcome the high surface reactivity, Ge/SiC$_x$N$_y$ core-shell structures were obtained in one-step growth by CVD of a germanium oxide precursor, [Ge{N(SiMe$_3$)$_2$}$_2$].[19] The decomposition pattern of the molecular precursor was determined by mass spectrometry to be as followed (eq. 2):

$$[Ge\{N(SiMe_3)_2\}_2] \rightarrow Ge^+ + NSi_xMe_y^+ + H_2^+ \tag{2}$$

Morphology and microstructure analyses (Fig. 2a) confirmed a catalyst-free growth and core-shell configuration where single crystalline Ge core was covered by an amorphous SiC$_x$N$_y$ overlayer. The HR-TEM image (Fig. 2b) of individual wire revealed single crystalline Ge nanowire with *fcc* lattice structure and a preferred growth direction of [11-2].

In the oxidation experiments, phase evaluation of Ge/SiC$_x$N$_y$ core-shell structures and Ge nanowire systems monitored by surface-sensitive X-ray photoemission spectroscopy showed the appearance of Ge 3d spectral features at the higher binding energy corresponding to oxidized Ge species in pure Ge NWs, whereas the Ge^0 phase in Ge/SiC$_x$N$_y$ samples showed no change before and after heat-treatment (Fig. 3). Given the fact that SiC$_x$N$_y$ overlayers are can efficiently passivate the Ge core, the Ge/SiC$_x$N$_y$ nanowires exhibit promising potential for device applications, especially due to the unique structural and electrical properties.

Figure 2. SEM and TEM images of Ge/SiC$_x$N$_y$ core-shell structure.

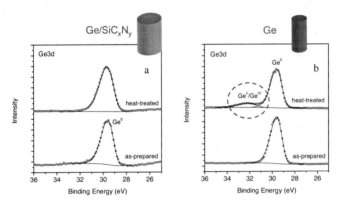

Figure 3. The corresponding XPS spectra of Ge 3d peaks of Ge/SiC$_x$N$_y$ (a) and Ge (b) nanowires.

Metal oxide nanowires are usually grown by high temperature vapor transport reactions, which lead to nanowires with different diameters, lengths and growth directions not desirable for device applications. Tin oxide nanowires have been synthesized by molecule-based CVD at relatively low temperature. Tin oxide is a wide-band semiconductor material (3.6 eV at 300K) that is transparent in the visible light region and useful as conductive electrodes[21] and treated as proto- type material for metal oxide sensors (MOS). We have grown tin oxide nanowires on gold-coated alumina substrates by decomposition of a volatile metal organic precursor, [Sn(O^{t}Bu)$_4$] (Fig. 4a).[20] The diameters of the obtained SnO$_2$ NWs ranged from 30-60 nm and the length was typically found to be in the range

15-20 μm. The TEM image revealed the single crystalline feature of SnO_2 NWs (Fig. 4b). The major gas phase species liberated in the decomposition of $[Sn(O^tBu)_4]$ were found to be tert-butyl alcohol and iso-butene by mass spectrometry analysis (eq. 3).

$$[Sn(O^tBu)_4] \rightarrow SnO_2 + i\text{-}C_4H_8 + t\text{-}BuOH \qquad (3)$$

Under constant precursor flux, growth of one-dimensional structures was achieved in the range 550-800°C whereas formation of granular films was observed at lower temperature ($<$ 525°C). The observation that anisotropic growth dominates at high temperature ($>$ 550°C) can be explained by the fact that no alloying between Au and Sn occurs below 550°C, which is an important requisite for the growth of nanowires.[20]

Figure 4. SEM and TEM images of SnO_2 nanowires.

Since the crystallographic orientation of the substrate influences the degree of alignment and control over the growth direction, ordered SnO_2 nanowires were obtained on single crystalline $TiO_2(001)$, (100) and (110) substrates as shown in Figure 5 (top view), which revealed that quite different growth orientations are observed under similar experimental growth conditions (750°C).

We have used as-grown SnO_2 nanowires as "substrates" to deposit different materials to obtain core-shell structures, such as $SnO_2@TiO_2$, $SnO_2@Fe_3O_4$ and $SnO_2@Ni$. The synthesis of heterostructure is motivated by the expectation to create new interfaces which are precisely defined and can allow lattice-level modifications of materials both in doped and composite materials. The gas sensing performance of SnO_2 based nanostructures in 90 ppm ethanol atmosphere at 250°C is shown in Figure 6. The sensitivity (R_a/R_g, where R_a and R_g defined as the resistance in dry air and ethanol gas, respectively) increases obviously from 1.6 to 2.2 till 3.1. It means that the sensitivity of modified SnO_2 nanostrucutres is strongly influenced by surface modification.

Figure 5. SEM images of SnO$_2$ nanowire arrays obtained on TiO$_2$ substrates with different orientations.

Further, SnO$_2$@SnO$_2$ and SnO$_2$@VO$_x$ heterostructures were obtained on SnO$_2$ nanowires by sequential decomposition of [Sn(O^{t}Bu)$_4$] and [VO(O^{i}Pr)$_3$], respectively (Fig. 7). The morphology of SnO$_2$@SnO$_2$ is typical brush-type heterostructures whereas SnO$_2$@VO$_x$ looks like VO$_x$ nanoparticles anchored on the surface of SnO$_2$ nanowires. The microstructure can be turned by controlling the flow rates of the two precursors. Ultra thin SnO$_2$ NWs were obtained after second-step growth. The diameter of SnO$_2$ NWs obtained as branches is approximately 10 nm (Inset, Fig. 7a), whereas, the diameter of first-stage nanowires is at least 20 nm.

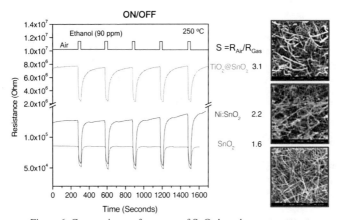

Figure 6. Gas sensing performance of SnO$_2$ based nanostructures.

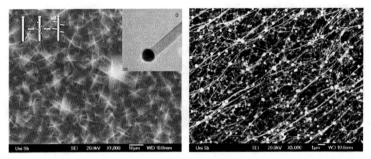

Figure 7. SEM images of (a) $SnO_2@SnO_2$ and (b) $SnO_2@VO_x$ heterostructures.

The surface properties of heterostructures $SnO_2@SnO_2$ and $SnO_2@VO_x$ compared with that of the original SnO_2 nanowires are studied by contact angle measurement (1 µL water) (Fig. 8). The contact angle of SnO_2, $SnO_2@VO_x$ and $SnO_2@SnO_2$ varied from 6.3°, 117.1° to 133.3°, respectively, indicating that the surface property changed from hydrophilic to higher hydrophobic possibly due to the increasing roughness and the presence of hyper-branched structure, which modify the wetting properties of the surface as known for the lotus leaf.

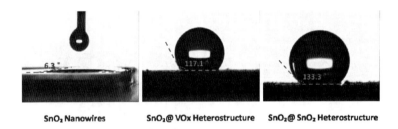

SnO₂ Nanowires   SnO₂@ VOx Heterostructure   SnO₂@ SnO₂ Heterostructure

Figure 8. Contact angle measurement on the surfaces of different nanostructures.

One-dimensional semiconductor nanowires represent several unique properties, important for functional behaviors, however, their device applications are still in an early stage and several technical issues need to be addressed before their potentials can be realized for industrial applications.[22] For instance, the chemical, thermal, mechanical stability of 1D nanostructures still need to be systematically studied in relation to their diameters. Concepts for the assembly of nanowires into complex structures or device architectures are also a challenge.

CONCLUSION

We described herein a low-temperature molecule-based CVD approach for controlled growth of 1D semiconductor nanostructures. Combining V-L-S catalytic growth mode with single molecular sources allowed precise control over dimensions, growth directions and surface states in nanowires. Low temperature and clean stripping of chemical designed precursors resulted in high purity of nanowires. Further a two-step growth strategy was developed to synthesize one dimensional $SnO_2@SnO_2$ and $SnO_2@VO_x$ heterostructures. Such 1D heterostructures can play an important role in the device applications due to structure and surface modification (Scheme 3). Although the molecule-process-structure-property relationship is complicated, the 1D nanostructures by MB-CVD showed better structure and surface regulation due to low synthesis temperatures and controlled decomposition behaviors of molecular sources.

Scheme 3. Roadmap of nanowire based devices.

REFERENCES

1. Z. L. Wang (ed.), *Nanowires and Nanobelts-Materials, Properties and Devices*, Kluwer Academic Publisher, Boston (2003).
2. X. Duan and C. M. Lieber, General Synthesis of Compound Semiconductor Nanowires, *Adv. Mater.*, **12**, 298-302 (2001).
3. M. S. Gudiksen, J. Wang, and C. M. Lieber, Synthetic Control of the Diameter and Length of Single Crystal Semiconductor Nanowires, *J. Phys. Chem. B*, **105**, 4062-64 (2001).
4. Y. Cui, X. Duan, J. Hu, and C. M. Lieber, Doping and Electrical Transport in Silicon Nanowires, *J. Phys. Chem. B*, **104**, 5213-16 (2000).
5. X. Duan, Y. Huang, Y. Cui, J. Wang and C. M. Lieber, Indium Phosphide Nanowires as Building Blocks for Nanoscale Electronicand Optoelectronic Devices, *Nature*, **409**, 66-69 (2001).
6. Y. Cui, and C. M. Lieber, Functional Nanoscale Electronic Devices Assembled Using Silicon Nanowire Building Blocks, *Science*, **291**, 851-54 (2001).
7. Y. Huang, X. Duan, Y. Cui, L. Lauhon, K. Kim and C. M. Lieber, Logic Gates and Computation from Assembled Nanowire Building Blocks, *Science*, **294**, 1313-18 (2001).
8. A. M. Morales and C. M. Lieber, A Laser Ablation Method for the Synthesis of Crystalline Semiconductor Nanowires, *Science*, **279**, 208-12 (1998).
9. Y. Wu and P. D. Yang, Germanium nanowire growth via Simple Vapor Transport, *Chem. Mater.*, **12**, 605-07 (2000).

10. J. R. Heath and F. K. Legoues, A Liquid Solution Synthesis of Single-Crystal Germanium Quantum Wires, *Chem.Phys. Lett.*, **208**, 263-68 (1993).

11. K. K. Likharev, and T. Claeson, Single-Electronics, *Sci. Am.*, **80**,1-9 (1992).

12. C. Schönenberger, B. M. I. Van der Zande, L. G. J. Fokkink, M. Henny, C. Schmid, M. Krüger, A. Bachtold, R. Huber, H. Birk, and U. Staufer, Template Synthesis of Nanowires in Porous Polycarbonate Membranes: Electrochemistry and Morphology, *J. Phys. Chem. B*, **101**, 5497-5505 (1997).

13. H. Adhikari, A. F. Marshall, C. E. D. Chidsey, and P. C. McIntyre, Germanium Nanowire Epitaxy: Shape and Orientation Control, *Nano Lett.*, **6**, 318-23 (2006).

14. S. Kodambaka, J. Tersoff, M. C. Reuter, and F. M. Ross, Germanium Nanowire Growth below the Eutectic Temperature, *Science*, **316**, 729-32 (2007).

15. K. L. Choy, Chemical Vapour Deposition of Coatings, *Progress in Materials Science*, **48**, 57-170 (2003).

16. R. S. Wagner, *Whisker Technology*, Wiley, New York (1970).

17. S. Mathur, M. Veith, T. Ruegamer, E. Hemmer and H. Shen, Chemical Vapour Deposition of MgAl$_2$O$_4$ Thin Films Using Different Mg-Al Alkoxides: Role of Precursor Chemistry, *Chem. Mater.*, **16**, 1304-12 (2004).

18. S. Mathur, H. Shen, V. Sivakov and U. Werner, Germanium Nanowires and Core-Shell Nanostructures by Chemical Vapor Deposition of [Ge(C$_5$H$_5$)$_2$], *Chem. Mater.*, **16**, 2449-56 (2004).

19. S. Mathur, H. Shen, N. Donia, T. Rügamer, V. Sivakov, and U. Werner, One-Step Chemical Vapor Growth of Ge/SiC$_x$N$_y$ Nanocables, *J. Am. Chem. Soc.*, **129**, 9746-52 (2007).

20. (a) S. Mathur, S. Barth, H. Shen, J. C. Pyun, and U. Werner, Size-Dependent Photo-Conductance in SnO$_2$ Nanowires, *Small*, **1**, 713-17 (2005). (b) S. Mathur and S. Barth, Molecule-based Growth of Aligned SnO$_2$ Nanowires and Branched SnO$_2$/V$_2$O$_5$ Heterostructures, *Small*, **3**, 2070-75 (2007).

21. P. G. Harrison and M. J. Willet, the Mechanism of Operation of Tin (IV) Oxide Carbon-Monoxide Sensors, *Nature*, **332**, 337-39 (1998).

22. Y. N. Xia, P. D. Yang, Y. G. Sun, Y. Y. Wu, B. Mayers, B. Gates, Y. D. Yin, F. Kim and H. Q. Yan, One-Dimensional Nanostructures: Synthesis, Characterization and applications, *Adv. Mater.*, **15**, 353-89 (2003).

NANOWIRES AS BUILDING BLOCKS OF NEW DEVICES: PRESENT STATE AND PROSPECTS

F. Hernandez-Ramirez[a,b,*], J. D. Prades[a], R. Jimenez-Diaz[b], S. Barth[c], A. Cirera[b], A. Romano-Rodriguez[b], S. Mathur[c], J. R. Morante[a,b]

[a] Institut de Recerca en Energia de Catalunya (IREC), Barcelona, Spain
[b] IN²UB/XaRMAE, Departament d'Electrònica, Universitat de Barcelona, Barcelona, Spain
[c] Department of Inorganic Chemistry, University of Cologne, Germany

ABSTRACT
    Nanowires have emerged as potential building blocks of new circuit architectures and innovative devices because of their properties, which are directly related to their reduced size and well-controlled chemical and physical properties. To date, intensive research efforts are carried out to attain a complete control of their synthesis and electrical performance. However, the fabrication of real and market-oriented nanodevices remains in the preliminary stage of development. In this work, the integration of individual metal oxide nanowires and consumer-class portable electronics is presented, and the use of the resulting systems to obtain proof-of-concept devices is shown. These systems have demonstrated their suitability to be used as gas sensors and ultraviolet (UV) photodetectors with high sensitivity, stable and reproducible responses; making possible the study of the phenomena which are characteristic at the nanoscale.

INTRODUCTION

    Crystalline semiconductor nanowires usually display interesting features which are directly related to their unique properties, such as the high surface-to-volume-ratio and well-defined atomic arrangement [1]. This makes them excellent candidates to be used as building-blocks of innovative devices (i.e. gas sensors, photodetectors, etc. [1-4]) in terms of stability, integration, robustness and packaging [5]; since most of the technological drawbacks which are commonly found in their micro and macro counterparts, such as the high power consumption, are easily circumvented [6]. Nevertheless, the development of competitive devices based on nanomaterials is still in the preliminary stages, and the launch of commercial products remains as a major and unsolved challenge due to the difficulties in electrically contacting nanowires and making the best use of their full potential [6]. Here, the current status of development of devices based on individual metal oxide nanowires is surveyed, and the main technological challenges which act as bottleneck to their potential use in real applications are presented.

ON THE ADVANTAGES OF INDIVIDUAL NANOWIRE-BASED DEVICES

    Metal-oxide devices are based on the chemico-electrical transduction reactions which take place at their surfaces [4, 7]. Thus, increasing the surface-to-volume ratio is the best strategy to maximize their response towards different external stimuli such as gas molecules or impinging UV photons. Individual single-crystalline nanowires meet this condition, and thanks to their low mass they are excellent candidates to be integrated in a new generation of low-consumption devices [8]. For instance, metal oxide nanowires can be heated up to the optimal operating temperature for gas sensing applications by probing current flowing through them with extremely low power consumption due to their small mass, giving rise to devices more efficient than their nanoparticle-based counterparts [6].
    Nevertheless, the controlled manipulation of individual nanowires is by no means a straightforward process, and makes necessary well-controlled and advanced nanofabrication

techniques [9]. To circumvent this technological obstacle, most of the overwhelming number of works dealing with metal oxide nanowires carried out so far is based on the use of bundles of nanowires instead of individual ones. However, the typical problems of thin-film metal-oxide devices are found in the characterization of multiple nanowires devices as well (i.e. parasitic electrical contributions and stability problems).

Figure 1a shows an sketch of a commercial metal oxide sensor formed by a layer of nanoparticles, whose temperature is modulated to activate gas reactions at their surfaces [4,10]. The operating principle of these sensors is based on their conductance modulation under exposure to gas. In a rough approximation, grain boundaries are described by highly resistive barriers, which represent the main contribution to the overall device resistance (fig.1a). Thus, the modulation of barriers among nanoparticles constitutes the most important gas transduction mechanism in porous-film sensors [10,11]. However, the random aggregation of nanoparticles in their inside as well as the spread in size makes difficult the study of the gas transduction phenomena. The particular geometry of each neck and boundary, and the unique orientation of the adsorbate-modulated electrical field regarding the externally applied bias field present in conductometric measurements make necessary to describe thin-film sensors with simplified models, and to analyze their responses as the convolution of diverse contributions [11]. The same conclusions are found if bundles of nanowires are used instead of nanoparticles, since randomly oriented boundary effects are not eliminated either (fig.1b).

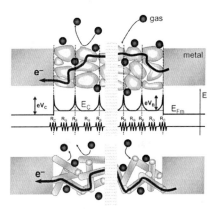

Fig. 1 Sketchs of two different types of conductometric gas sensors based on metal oxides. (a) Commercial thin-film sensor formed by nanoparticles. Here, electrons must go through a network of nanocrystals with different size and shape. From an energy point of view, electrons are to overcome potential barriers [(i) metal-semiconductor barriers (e_{VC}) and (ii) intergrain boundary barriers (e_{VB})]. The influence of gas on the height of the barriers determines the final response of the sensor. This is equivalent to a network of resistors [(i) metal-semiconductor contacts (R_C), (ii) grain boundary interfaces (R_B) and (iii) metal-oxide grains (R_G)]. (b) Multi-nanowire sensor. The above mentioned discussion is valid here as well

To contrast, the abovementioned contributions are circumvented if individual nanowire-based sensors are studied (Fig.2). In this experimental scenario, direct responses towards gases are monitored, because gas diffusion among nanograins / nanowires is eliminated. Thus, the sensing principle of individual metal oxide nanowires can be theoretically described by pure surface effects [5]. Adsorption of gas molecules at the nanowire modulates the width of a depleted region close to the external shell. This effect based on the capture and release of electrical charges at the surface modifies the conduction channel through the nanowire and as a consequence the electrical resistance R_{NW} (Fig.2), providing direct and fundamental information of the electrical charge exchange and the role of surface states n_v in the sensing process [12]. According to this model, R_{NW} under exposure to gas is given by:

$$R_{NW} = \frac{\rho L}{\pi (r - \lambda)^2} \qquad (1)$$

where ρ is the nanowire resistivity, L the nanowire's length, r the nanowire's radius and λ the width of the depletion layer created by adsorbed molecules. Equation 1 lays down a connection between the nanowire's radius and the gas response: the thinner the nanowire is, the higher the gas response [12].

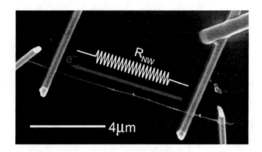

Fig.2 Single-nanowire sensor. If a nanowire is measured in 4-probe DC configuration, the conductometric response is basically determined by changes of the conduction channel along the nanowire (R_{NW}). On the contrary, contacts effects are overcome.

Nanowires with radii below r = 40 nm are commonly used to maximize the response of the resultant gas sensors, giving rise to technological issues derived from working at the nanoscale such as high contact resistance at the metal-nanowire interfaces [13]. Nevertheless, most of these problems are circumvented with different operating strategies reported elsewhere [8, 13], paving the way to better gas nanosensors than their thin-film counterparts (Fig.3).

On the other hand and as far as the use of nanowires as photodetectors is concerned, it was demonstrated elsewhere [14] that individual SnO_2 and ZnO nanowires exhibit outstanding response towards impinging UV photons (Fig 4). For this reason, systematic fabrication strategies to enhance their responses are being evaluated [14].

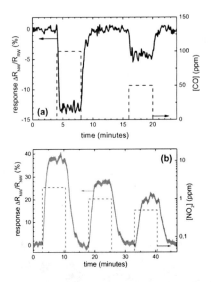

Fig. 3 (a) Response of a tin oxide nanowire towards different CO pulses at T = 300°C. (b) Response of a tin oxide nanowire towards NO_2 pulses at T = 175°C. In both cases, fast and reproducible behaviors are monitored, with excellent recovery of the synthetic air resistance baseline

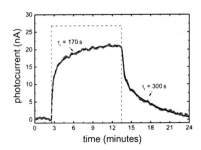

Fig. 4 Dynamic behaviour of the photoresponse I_{ph} measured with a single ZnO nanowire when a UV pulse is applied (dashed line) ($\Phi_{ph} = 3.3 \cdot 10^{18}$ $ph\,m^{-2}s^{-1}$; $\lambda = 340 \pm 10\,nm$; $V = 1V$). Here, Φ_{ph} is the photon flux, λ is the photon wavelenght and V is the bias voltage applied during the measurements.

In short, it should be pointed out that single-nanowire prototypes are extremely useful to demonstrate the potential of nanomaterials. Nevertheless, results and responses obtained so far can only be improved if complex architectures, such as multiple-nanowires contacted in parallel, are fabricated and characterized. For this reason, many research efforts are currently been devoted to reach this ambitious goal.

TOWARDS REAL DEVICES

Monitoring the electrical properties of individual nanowires using portable and affordable consumer-class electronics (Fig.5) was recently demonstrated [15]. These devices, compared to lab equipments, were able to detect and quantify the response of individual nanowires towards UV light pulses and different gases with long-term stability thanks to the low current injected by the platform to the nanowire [15].

Fig.5 Low-cost electronic platform designed to characterize nanowires

It is well-known that metal oxide materials need to be heated at a specific temperature to maximize their response to a specific target [11]. Therefore, the use of a heater becomes a necessary tool to modulate the final performance of these materials. To solve this issue, both bottom-up and top-down fabrication techniques have been successfully integrated in a single process; nanowires are electrically contacted to a micro-hotplate with an integrated heater [15] (Fig.6).

Fig.6 (right) Individual SnO$_2$ nanowire electrically contacted to two platinum electrodes. (left) Interdigitated platinum electrodes surrounded by a microheater. The nanowire is located in the middle (red square)

This set up allows modulating the effective temperature of the wire as function of the power dissipated at the heater in a fast and completely reproducible way. It is noteworthy that this solution combined with a good electronic interface, which integrates the thermal control of the nanowire, is extremely useful in many sensing applications [15]. Other technological alternatives are also being explored. A typical example is the design self-heated nanowires-devices to eliminate the use of external heaters [6]. Although these studies are currently ongoing, extremely promising results were obtained so far [8].

FUTURE CHALLENGES

The need to obtain good electrical contacts in a controlled and reproducible process has forced to look for innovative nanofabrication approaches. Nowadays, metal stripes with well-defined shapes in the nanometer range and high electrical quality are easily fabricated with different techniques such as Focused Ion Beam (FIB-), e-beam- or UV- and shadow-mask-lithography, enabling the fast engineering of advanced proof-of-concept devices. However, most of these techniques and particularly the nanofabrication protocols based on them evaluated so far are only suitable for research prototyping, since they are not scalable, and therefore do not fulfil the requirements to take a leap in industry.

For this reason, new solutions to solve the problem of scalability of nanowire-based devices and to edit the first complex circuit with them are currently under evaluation, such as self-assembly approaches [16], and the use of other techniques like electrospinnig and microcontact / ink-jet printing [17-21]. Besides enabling the integration of nanowires in industrial processes, the chosen solution should palliate the poor reproducibility of present proof-of-concept devices (compared to the standards of the microelectronic industry), which is mainly originated by the lack of well-established fabrication and characterization methodologies. However, it should be highlighted that all these fabrication techniques are still in a preliminary stage of development despite some promising results were recently reported [22].

In short, nanowire-based devices face up to some issues to be transferred to industrial production, but they also exhibit some intrinsic advantages compared to their micro and macro counterparts.

CONCLUSION

Crystalline nanowires have novel properties derived from their reduced dimensions and excellent atomic arrangement, which can be used to obtain functional devices such as gas sensors and UV photodetectors better than their bulk-counterparts. Up to now, simple prototypes based on few nanowires have been fabricated and studied. Nevertheless, complex device architectures remain as unattainable objective due to the difficulties of working at the nanoscale in a controlled way. For this reason, self-assembly techniques (i.e. dielectrophoresis) or other alternatives such as electrospinning are considered excellent fabrication alternative to overcome this technological bottleneck and thus, to develop commercial devices with nanowires in the future.

FOOTNOTES
* Corresponding author. E-mail: fhernandezra@gmail.com

REFERENCES

[1] M. Law, J. Goldberger and P. Yang, Annu. Rev. Mater. Res., **2004**, 34, 83.

[2] S. V. N. T. Kuchibhatla, A. S. Karaoti, D. Bera and S. Seal, Progress Mater. Sci., **2007**, 52, 699.

[3] G. Eranna, B. C. Joshi, D. P. Runthala and R. P. Gupta, Critic. Rev. Sol. State Mater. Sci., **2004**, 29, 188.

[4] G. Korotcenkov, Sens. Actuators B: Chem., **2005**, 107, 209

[5] E. Comini, C. Baratto, G. Faglia, M. Ferroni, A. Vomiero, G. Sberveglieri, Progress in Mater. Sci., **2009**, 54, 1

[6] F. Hernandez-Ramirez, J. D. Prades, R. Jimenez-Diaz, T. Fischer, A. Romano-Rodriguez, S. Mathur, J. R. Morante. Phys. Chem. Chem. Phys., **2009**, DOI: 10.1039/B905234H.

[7] A. Kolmakov, D. O. Klenov, Y. Lilach, S. Stemmer and M. Moskovits, Nano Lett., **2005**, 5, 667.

[8] J. D. Prades, R. Jimenez-Diaz , F. Hernandez-Ramirez, S. Barth, J. Pan, A. Cirera, A. Romano-Rodriguez, S. Mathur and J. R. Morante, Appl. Phys. Lett., **2008**, 93, 123110.

[9] F. Hernandez-Ramirez, A. Tarancon, O. Casals, J. Rodríguez, A. Romano-Rodriguez, J. R. Morante, S. Barth, S. Mathur, T. Y. Choi, D. Poulikakos, V. Callegari and P. M. Nellen, Nanotechnol., **2006**, 17, 5577.

[10] G. Korotchenkov, Mater. Sci. Engin. R, **2008**, 61, 1.

[11] N. Barsan, D. Koziej and U. Weimar, Sens. Actuators B: Chem., **2007**, 121, 18.

[12] F. Hernandez-Ramirez, J. D. Prades, A. Tarancon., S. Barth, O. Casals, R. Jimenez-Diaz, E. Pellicer, J. Rodriguez, J. R. Morante, M. A. Juli, S. Mathur and A. Romano-Rodriguez, Adv. Funct. Mater., **2008**, 18, 2990.

[13] F. Hernandez-Ramirez, A. Tarancon, O. Casals, E. Pellicer, J. Rodriguez, A. Romano-Rodriguez, J. R. Morante, S. Barth, Mathur. Phys. Rev. B, **2007**, 76, 085429.

[14] J. D. Prades, R. Jimenez-Diaz, F. Hernandez-Ramirez, L. Fernandez-Romero, T. Andreu, A. Cirera, A. Romano-Rodriguez, A. Cornet, J. R. Morante, S. Barth and S. Mathur, J. Phys. Chem. C, **2008**, 112, 14639.

[15] F. Hernandez-Ramirez, J. D. Prades, A. Tarancon, S. Barth, O. Casals, R. Jimenez-Diaz, E. Pellicer, J. Rodríguez, M. A. Juli, A. Romano-Rodriguez, J. R. Morante, S. Mathur, A. Helwig, J. Spannhake and G. Mueller, Nanotechnol., **2007**, 18, 495501.

[16] S.Kumar, S. Rajaraman, R.A. Gerhardt, Z.L. Wang, P.J. Hesketh. Electrochim. Acta, **2005**, 51, 943.

[17] D. Lin, H. Wu, R. Zhang, W. Pan. Nanotechnology, **2007**, 18, 465301.

[18] K.M. Sawicka, A.K. Prasad, P.I. Gouma. Sensors Lett., **2005**, 3, 31.

[19] T.H.J. van Osch, J. Perelaer, A.W.M. de Laat, U.S. Schubert. Adv. Mater., **2008**, 20, 343.

[20] Y.-K. Kim, S.J. Park, J.P. Koo, D.-J.Oh, G.T. Kim, S. Hong, J.S. Ha. Nanotechnology, **2006**, 17, 1375.

[21] J.-W. Song, J. Kim, Y.-H. Yoon, B.-S. Choi, J.-H. Kim, C.-S. Han. Nanotechnology, **2008**, 19, 095702.

[22] Z. Fan, J.C. Ho, Z.A. Jacobson, H. Razavi, A. Javey. PNAS, **2008**, 105, 11066.

PREPARATION OF TiO$_2$-NANOPARTICLES-THIN FILMS BY ELECTROPHORESIS DEPOSITION METHOD

Kazuatsu Ito, Yuuki Sato, Motonari Adachi, Shinzo Yoshikado
Department of Electronics, Doshisha University,
Kyotanabe 610-0321, Japan

ABSTRACT

Titanium oxide (TiO$_2$)-nanoparticles of anatase crystal structure were synthesized and deposited using the electrophoresis deposition method. All of TiO$_2$-nanoparticles were charged positively in water or ethanol colloidal solution, and were attracted toward indium-tin oxide (ITO) glass substrate as the negative electrode as well as TiO$_2$-particles (P25) of which diameter is larger than that of TiO$_2$- nanoparticles. Large cracks were generated in the TiO$_2$-thin film and the thin film was flaked off from the substrate with increasing the deposition time in water. This is because amounts of TiO$_2$-nanoparticle transported by electrophoresis decreased with increasing the deposition time, the density of TiO$_2$- nanoparticles-thin film became low, and the thin film contracted during dry process. The cracks were reduced using ethanol as a solvent compared to using water as a solvent. Meanwhile, when the deposition time was short, TiO$_2$-nanoparticles-thin film had high optical transparency and no apparent cracks. Moreover, the thickness of TiO$_2$-nanoparticles-thin film, which has high optical transparency and no apparent cracks, could be increased by repeating the short time deposition in ethanol.

INTRODUCTION

Anatase crystal structure titanium oxide (TiO$_2$) has been focused as a photo-catalyst[1]. TiO$_2$ also have been studied actively as material for electrode of dye-sensitized solar cells. The quality of TiO$_2$-thin film is one of the most important factors determining the characteristics of the dye-sensitized solar cell[2]. For the above uses, TiO$_2$ is used in the form of thin film. In the case of a photo-catalyst, small sized TiO$_2$-particles are required to enlarge the surface area that comes in contact with a reactant. In addition, TiO$_2$-thin film must have high optical transparency so that the incident light passes through inside the thin film to generate holes and electrons, and to act as a photo-catalyst. Furthermore, in the case of a dye-sensitized solar cell, it is required that thin film has no apparent cracks which result in the decrease of electromotive force and the thickness is in the range from 10 to 15 µm. It has been reported that TiO$_2$-thin film with high optical transparency can be deposited using plasma deposition method, such as radio frequency (RF) magnetron sputtering method[1]. However, in the case of plasma deposition, the deposition rate of is low, the substrate heating is required, and the production cost is high. On the other hand, electrophoresis deposition method is low-cost process and the deposition rate of electrophoresis is high[3]. In addition, electrophoresis deposition method enables the deposition of thin film on the conductive substrate having the arbitrary area or the arbitrary shape. Furthermore, the deposition of the thin film without apparent crack is possible, the thickness of TiO$_2$-thin film can be easily controlled by adjusting electrophoresis time or current density, and the high reproducibility of quality is expected[3]. However, the size of TiO$_2$-particle used in electrophoresis deposition method should be around nanometer to prepare highly dispersed colloidal solution.

Although the particle size of TiO$_2$-nanoparticle is approximately 5~15 nm and smaller than that of TiO$_2$-particles such as P25 (average particle size of approximately 20 nm), deposition of a porous thin film is expected using TiO$_2$-nanoparticle because TiO$_2$-nanoparticle is composed of the connection of small sized TiO$_2$-particles. The crystal lattices between particles of TiO$_2$-nanoparticle are connected with each other by oriented-attachment-mechanism and one-dimensional structure is formed[4]. Therefore, electrons can be transported easily between particles. In this study,

TiO_2-nanoparticles were synthesized and deposited using the electrophoresis deposition method in various colloidal solutions to fabricate crack-free thin film with high optical transparency[5].

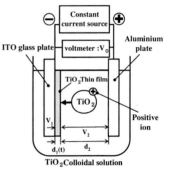

Figure 1. Schematic drawing of the constant current electrophoresis method.

ELECTROPHORESIS DEPOSITION

Electrophoresis is the phenomenon that high polymer molecules or colloidal particles charged in a solution move in the solution by applying an electric field. The interface movement phenomenon is determined by zeta (ζ)-potential, which is the electric potential difference between anchoring layer and the solution. TiO_2-particles are attracted toward negative electrode and is deposited on the metal or conductive layer such as indium-tin oxide on the glass substrate (ITO glass substrate). Thus, TiO_2-particles are deposited on the substrate and thin film is formed during the flow of electrophoresis current. For the electrophoresis deposition, two methods have been used; constant current electrophoresis method and constant voltage electrophoresis method. In the case of depositing TiO_2-thin film, it has been reported that thin film with high quality can be obtained using constant current electrophoresis method[3]. Therefore, in this study, the constant current electrophoresis method is used to deposit TiO_2-thin film. A schematic drawing of the constant current electrophoresis method is shown in figure 1. Relation of voltage between the two electrophoresis electrodes ($V_0(t)$) and thickness ($d_1(t)$) of thin film is given by[3]

$$V_0(t) = \frac{\rho_1}{S_0} d_1(t) I_0 + \frac{\rho_2}{S_0} d_2 I_0 . \tag{1}$$

Here, t is an electrophoresis time, S_0 is an area of electrode, ρ_1 is a resistivity of TiO_2-thin film, ρ_2 is a resistivity of colloidal solution, d_2 is the distance between electrodes ($d_2 \gg d_1$), I_0 is electrophoresis current. If one TiO_2-particle transports the positive charges of $n^+ e$ [C] for 1 s and only these charges contribute to I_0, $I_0/(n^+ e)$ TiO_2-particle are adsorbed on the ITO glass substrate for 1 s. When the particles are deposited without space, $S_0/\pi a^2$ TiO_2-particles are adsorbed for each one layer. Therefore, $d_1(t)$ is given by

$$d_1(t) = 2a \frac{I_0}{n^+ e} \frac{\pi a^2}{S_0} t = \frac{2\pi a^3 I_0}{n^+ e S_0} t . \tag{2}$$

Here, a is the radius of a TiO_2 particle and e is elementary charge. Thus, the number n^+ of positive charge adsorbed on a TiO_2-particle is given by

$$n^+ = \frac{2\pi a^3 I_0}{d_1(t) e S_0} t . \tag{3}$$

EXPERIMENT

Preparation of TiO_2-nanoparticles

Laurylamine hydrochloride (LAHC) was dissolved in distilled water and tetra-isopropylorthotitanate (TIPT) was mixed with acetylacetone (ACA) to decrease the hydrolysis and the condensation rates of TIPT. When TIPT was mixed with the same molar ratio of ACA, the color changed from colorless to yellow. This yellow TIPT solution was added to 0.1 M LAHC aqueous solution. The molar ratio of TIPT to LAHC was 4. When two solutions were mixed, precipitation occurred immediately. The precipitates were dissolved completely by stirring the solution for several days at 40°C, and the solution became transparent. The temperature was then changed to 80°C. After 3 days, the solution became a white gel with a thin yellow liquid layer. After being washed with 2-propanol and successive 10 min centrifugation (10000 rpm), the TiO$_2$-nanoparticles were separated. This washing operation was repeated several times.

Electrophoresis deposition

ITO glass substrate as the negative electrode and an aluminum plate as the positive electrode were placed with a distance of 25 mm perpendicular to the liquid level in the glass beaker. TiO$_2$-nanoparticles of 0.38 g were added to solvent of 60 mℓ in a glass beaker. As a solvent, ion-free water (resistivity above 10^7 Ωcm, pH 6~7, here after water), methanol, and ethanol, which have different permittivity and viscosity, the mixture of water and ethanol (volume ratio is 1:1), and acetone were used. The concentration of the TiO$_2$-nanoparticle-colloidal solution was approximately 0.08 mol/ℓ. This molar ratio is optimum condition of depositing TiO$_2$-thin film[3]. Then, the colloidal solution was degassed using ultrasonic cleaner. TiO$_2$-nanoparticles-thin film was deposited by constant current electrophoresis deposition method by changing electrophoresis time. The temperature of colloidal solution was approximately 26 °C.

Evaluation of TiO$_2$-nanoparticles and TiO$_2$-nanoparticles-thin film

TiO$_2$-nanoparticles mixed with a small amount of ethanol were dropped on the mesh substrate and were dried. Then, these TiO$_2$-nanoparticles were observed using transmission electron microscope (TEM) (JEOL, JEM-2100F, accelerating voltage of 200 kV). X-ray diffraction (XRD) patterns of TiO$_2$-nanoparticles and the thin film were measured by pasting TiO$_2$-particles on ITO glass substrate. TiO$_2$-nanoparticles and the film were also observed using optical microscope (Nikon, ME600, All-in-Focus), scanning secondary electron microscope (SEM) (JEOL, JSM-7500, accelerating voltage of 10 kV). The thickness of TiO$_2$-thin film was measured using atomic force microscope (AFM) (SII, Nanopics 2100).

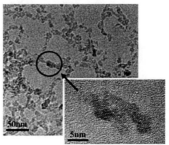

Figure 2. TEM image of TiO$_2$-nanoparticles.

RESULTS AND DISCUSSION

A TEM image of TiO$_2$-nanoparticles is shown in figure 2. Many of the TiO$_2$-nanoparticles were connected with each other like a chain and oriented-attachment-mechanism was also observed. As shown in figure 3, it was found from X-ray diffraction patterns that the crystal structure of TiO$_2$-nanoparticles (not shown in figure 3) and its thin film was anatase crystal structure.

TiO$_2$-nanoparticles or P25 were deposited using the constant current electrophoresis deposition method. All of TiO$_2$-nanoparticles were charged positively in aqueous colloidal solution, were

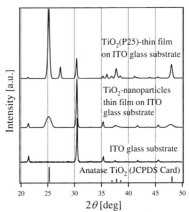

Figure 3. XRD patterns of TiO$_2$-thin films deposited using electrophoresis method.

attracted toward the negative electrode under electric field, and thin film was deposited on the ITO glass substrate as well as P25[3]. Electrophoresis current density and deposition time were 0.1 mA/cm^2 and 260 s, respectively. This current density is optimum one to deposit a high quality thin film[5]. Thin films were thermally annealed for 1 h at 450°C in air after deposition. Photographs, SEM images, and optical photomicrographs of TiO$_2$-nanoparticles and P25-thin films deposited under above conditions are shown in figures 4, 5, and 6, respectively. The thicknesses of thin films were approximately 1.8 μm for TiO$_2$-nanoparticle and approximately 9 μm for P25. P25-thin film of approximately 1 μm thicknesses is also shown in figure 4 (c) for the comparison. The transparency of TiO$_2$-nanoparticles-thin film was higher than that of P25-thin film. It is speculated that the low transparency of P25-thin film is owing to the scattering of light, because the particle size of P25 is larger than that of TiO$_2$-nanoparticle and P25 form aggregates as shown in figure 5 (b). It is found from figure 6 that the TiO$_2$-nanoparticles-thin film was separated like a scale and the number of cracks generated in TiO$_2$-nanoparticles- thin film was larger than that generated in P25-thin film. Moreover, interference stripes were observed in each flake. This is because the flake was peeled off like a scale and incidence light was interfered in the thin film.

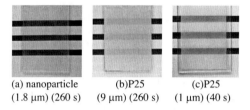

(a) nanoparticle   (b)P25   (c)P25
(1.8 μm) (260 s)   (9 μm) (260 s)   (1 μm) (40 s)

Figure 4. Photographs of thin films for nanoparticle and P25. Electrophoresis condition: 0.1 mA/cm^2 in water.

(a) nanoparticle (1.8 μm)   (b) P25 (9 μm)
Figure 5. SEM images of TiO$_2$-thin film deposited using (a) TiO$_2$-nanoparticles and (b) P25. Electrophoresis condition: 0.1 mA/cm^2, 260 s in water.

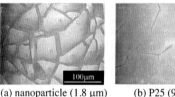

(a) nanoparticle (1.8 μm)   (b) P25 (9 μm)
Figure 6. Optical photomicrographs of TiO$_2$-thin films deposited using (a) TiO$_2$-nanoparticles and (b) P25. Electrophoresis condition: 0.1 mA/cm^2, 260 s.

Figure 7 shows the relation ($V_0(t)$-t characteristics) among the electrophoresis time t and the voltage $V_0(t)$ between the two electrophoresis electrodes. Electrophoresis current density and the deposition time were 0.1 mA/cm^2 and 300 s, respectively. In the case of P25, the $V_0(t)$ increased proportionally to the deposition time. Meanwhile, in the case of TiO$_2$-nanoparticles, the $V_0(t)$ increased rapidly until approximately 70 s (point A) and increased gradually after approximately 70 s. The reason for the increase in $V_0(t)$ is owing to the increase of thickness of thin film as given by equation (1). The generation of cracks is considered to be that the density of thin film becomes low, the water contained in the thin film evaporates and the thin film contracts. Therefore, it is speculated that the amount of transported TiO$_2$-nanoparticles changed at 70 s and it decreased after 70 s. This result indicates that mobile species other than positively charged TiO$_2$-nanoparticles contributed markedly to electrophoresis current after approximately 70 s because the constant current electrophoresis method was used. Therefore, to reduce mobile species other than TiO$_2$-nanoparticles, TiO$_2$-thin film were deposited in organic solvent of which ionization is small. Methanol, ethanol, acetone, and the mixture of water and ethanol (volume ratio is 1:1) were used as an organic solvent. The viscosity and relative permittivity of water, each alcohol or acetone at 25 °C are shown in table I. The ionization is high as the value of permittivity increases obeying to Coulomb's law.

When TiO$_2$-nanoparticles were added to acetone, TiO$_2$-nanoparticles were precipitated without dispersion. This result suggests that the ζ-potential of TiO$_2$-nanoparticles is almost zero in acetone. Meanwhile, when the solvent is other one except for acetone, TiO$_2$-nanoparticles were attracted toward ITO glass as the negative electrode. This result suggests that positive ions such as CH$_2$CH$_3^+$ in ethanol, and CH$_3^+$ in methanol adsorbed on TiO$_2$-nanoparticles.

TiO$_2$-nanoparticles were deposited using the electrophoresis deposition method with methanol, ethanol, and the mixture of water and ethanol. Electrophoresis current density and deposition time were 0.1 mA/cm^2 and 260s, respectively. $V_0(t)$-t characteristics of electrophoresis deposition are shown in figure 8, and value of the initial voltages $V_0(0)$ are shown in figure 9. $V_0(0)$ becomes small when the amount of mobile species such as OH$^-$ ions increases. Just after electrophoresis deposition, mobile species except for positively charged TiO$_2$-nanoparticle are attracted toward electrode immediately, then TiO$_2$-nanoparticles are attracted toward electrode gradually. When the incline of $V_0(t)$-t characteristics is small, it is speculated that the amount of mobile species, which don't contribute the electrification of TiO$_2$-nanoparticles, is large. When TiO$_2$-nanoparticles-thin films were deposited in methanol, the thickness of thin films were approximately 0.03 μm and the inclines of $V_0(t)$-t characteristics were much smaller than that deposited in other solvents. This result indicates that in methanol, the amount of transported TiO$_2$- nanoparticles is very small and the numbers of positive ions adsorbed on TiO$_2$-nanoparticles are smaller than that in other solvents.

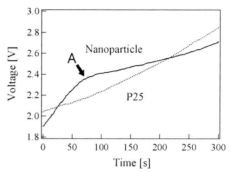

Figure 7. $V_0(t)$-t characteristics for electrophoresis method using TiO$_2$-nanoparticle and P25. Electrophoresis condition: 0.1 mA/cm^2, 260 s in water.

Table I. Viscosity and relative permittivity of solvent at 25 °C.

solvent	viscosity [mPa s]	relative permittivity
water	0.89	78.3
methanol	0.55	32.6
ethanol	1.09	24.3
acetone	0.30	20.7

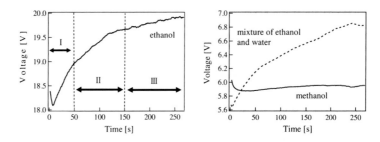

(a)   (b)

Figure 8. $V_0(t)$-t characteristics for electrophoresis method using TiO₂-nanoparticle in (a) methanol and mixture of ethanol and water and (b)ethanol. Electrophoresis condition:0.1 mA/cm², 260 s.

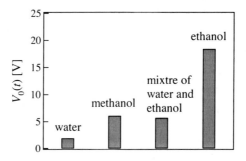

Figure 9. Voltage $V_0(0)$ at t=0 for each solvent.
Electrophoresis condition:0.1 mA/cm².

Optical photomicrographs of the thin film deposited in ethanol and the mixture of water and ethanol are shown in figure 10. Electrophoresis current density and deposition time were 0.1mA/cm^2 and 260 s, respectively. For comparison, optical photomicrograph of the thin film deposited in water is also shown in figure10 (a). When TiO$_2$-nanoparticles-thin film was deposited in the mixture of water and ethanol, electrophoresis deposition voltage increased from 5.6 V (V_0 (0)) to 6.8 V(V_0 (260)). However, a lot of cracks were observed in the thin film as well as that deposited in water. On the other hand, the thickness of TiO$_2$-nanoparticles-thin film deposited in ethanol was approximately 1.4 μm and the number of cracks generated in this thin film was much smaller than that generated in the film deposited in water or the mixture of water and ethanol. It is found from figure 9 that the initial voltage V_0 (0) for the mixture of water and ethanol is smaller than that for ethanol. The reason for the small value of V_0 (0) for the deposition in water is speculated to be that a lot of ions, such as H$_3$O$^+$ and OH$^-$, exist in water because the permittivity and the ionization degree for water are large, as shown in table I. Therefore, the density of TiO$_2$-nanoparticles thin film becomes low because the sizes of molecules such as H$_3$O$^+$ and OH$^-$ are small and the number of ions adsorbed on TiO$_2$-particles increases. On the other hand, the value of V_0 (0) for ethanol is approximately 9 times larger than that for water, and the ionization degree of ethanol is small. Thus, the number of positive ions such as CH$_2$CH$_3$$^+$ which are adsorbed on TiO$_2$-nanoparticles, and that of ions which contributes to electrification are small, and the electric field of electrophoresis increases. As a result, the amount of TiO$_2$-nanoparticles transported during electrophoresis in ethanol was larger than that in water. It is considered from above results that the thin film with high density could be deposited in ethanol and TiO$_2$-nanoparticles adhered to ITO glass substrate tenaciously. In conclusion, the cracks could be reduced and a thin film with high quality was obtained by changing solvent from water to ethanol. The cracks were reduced using ethanol as a solvent compared to using water as a solvent.

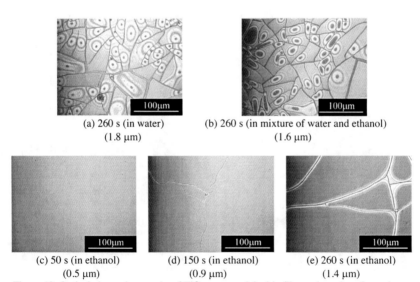

(a) 260 s (in water)   (b) 260 s (in mixture of water and ethanol)
(1.8 μm)   (1.6 μm)

(c) 50 s (in ethanol)   (d) 150 s (in ethanol)   (e) 260 s (in ethanol)
(0.5 μm)   (0.9 μm)   (1.4 μm)
Figure 10. Optical photomicrographs of TiO$_2$-nanoparticle thin films using ethanol as solvent. Electrophoresis condition:0.1 mA/cm^2.

When the electrophoresis current density is 0.1 mA/cm^2 and the electrophoresis time is 260 s in ethanol, it was found from figure 8 (b) that the $V_0(t)$ increased rapidly until approximately 50 s and then increased gradually after approximately 50 s. As shown in figure 11 (a), the amount of TiO$_2$-nanoparticles transported during electrophoresis is considered to be large until 50 s (region I shown in figure 11(a)). However, the number of positive ions adsorbed on TiO$_2$-nanoparticles varies with increasing deposition time after 50 s. In addition, three types of coordination for TiO$_2$-nanoparticles are also considered. In other words, although the thin film with high density was deposited until 50 s, the thin film with low density was deposited after 50 s (regions II and III shown in figure 11(a)) because a lot of positive ions are contained in the thin film. Therefore, the thin film deposited until 50 s hardly contracted and has no apparent cracks after dry process as shown in figure 10 (c) because the density of the thin film is high. On the other hand, some of cracks were observed in the thin film deposited for 150 and 260 s as shown in figure 10 (d), (e). Consequently, the thin film deposited for more than 50 s is contracted during dry process and the size of crack generated upper side of thin film became large as shown in figure 11(b). It is confirmed that bottom face of the crack is crack-free TiO$_2$-nanoparticles-thin film from SEM image. The relation between the incline of $V_0(t)$-t characteristics shown in figure 8 and the width of crack is shown in figure 12. The width of crack became large as the incline of $V_0(t)$-t characteristics decreases. The thickness of the thin film deposited by stopping electrophoresis deposition at 50 s and 150 s were approximately 0.5 μm and 0.9 μm, respectively. The relation between the incline of $V_0(t)$-t characteristics and the growth rate of thin film in water and ethanol is shown in table II. The growth rate of thin film was high when the incline of $V_0(t)$-t characteristics is large. It is speculated that TiO$_2$-nanoparticles with few positive ion are

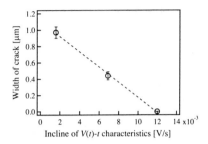

Figure 12. Relation between the incline of $V_0(t)$-t characteristics and width of cracks.

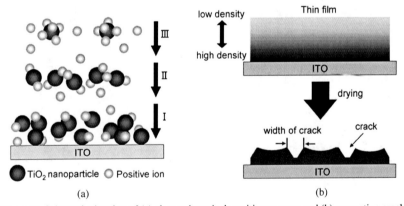

Figure 11. Schematic drawing of (a) electrophoresis deposition process and (b) generating cracks.

electrophoresed in the large incline range of $V_0(t)$-t characteristics. However, in the small incline range of $V_0(t)$-t characteristics, the number of positive ions adsorbed on TiO$_2$-nanoparticles increases and the electrophoresis of TiO$_2$-nanoparticles is reduced. Therefore, in order to increase the thickness of crack-free TiO$_2$-thin film with high density in ethanol, electrophoresis deposition was stopped at 50 s. After the deposition, TiO$_2$-nanoparticles-thin film was dried for more than 20 min in air. For drying time of 5 min, cracks with large width were generated all over the thin film at 4th time depositing. Then, TiO$_2$-nanoparticles were deposited on the dried TiO$_2$-nanoparticles-thin film under the same condition again in the colloidal solution stirred enough. This operation was repeated several times.

Relation between the thickness of TiO$_2$-nanoparticles-thin film and the repeated times of deposition is shown in figure 13. The thickness of thin film increased proportionally to the repeated times of deposition. When the deposition was repeated 7 times, the thickness of thin film became approximately 1.9 μm. Optical photomicrographs of TiO$_2$-nanoparticle-thin films with the repetition of 3 or 7 times deposition are shown in figure 14. Although cracks were generated at the edge of the thin film, at which strength of electric field is weak, thin film without apparent cracks could be deposited in the center of the thin film. $V_0(t)$-t characteristics between 1st and 3rd times deposition are shown in figure 15 (a) and those between 4th and 7th times deposition are shown in figure 15 (b). The incline of $V_0(t)$-t characteristics between 1st and 3rd times deposition is approximately 0.011 V/s and the initial voltage $V_0(0)$ increased as the thickness of thin film increased. On the other hand, the incline of $V_0(t)$-t characteristics at 4th times deposition was 0.008 V/s. This value is smaller than that obtained in the deposition between 1st and 3rd times deposition. Moreover, $V_0(t)$ between the two electrodes decreased at 6th times deposition. The reason for the decrease in $V_0(t)$ is considered that the drying of thin film is not enough as the thickness of films increases. Photograph of the thin film with the repetition of 7 times deposition is shown in figure 16 (a). For comparison, TiO$_2$-nanoparticle-thin film deposited in water under the condition of electrophoresis current density of 0.1 mA/cm^2 and electrophoresis time of 260 s and P25-thin film of approximately 1 μm thicknesses are also shown in figure 16 (b) and (c). It was

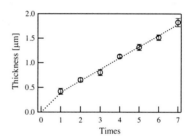

Figure 13. Relation between thickness of TiO$_2$-nanoparticle thin films and repeated times of deposition. Electrophoresis condition: 0.1 mA/cm^2, 50 s.

Table II. Relation between incline of $V(t)$-t characteristics and growth rate of thin films. Electrophoresis condition: 0.1 mA/cm^2.

solvent	region shown in Figs.9,10	deposition time [s]	incline of $V(t)$-t characteristics [V/s]	growth rate of thin films [nm/s]
water	I	0~50	0.007	10.0
	II	51~260	0.0018	6.2
ethanol	I	0~50	0.012	10.1
	II	51~150	0.0069	5.0
	III	151~260	0.0016	3.1

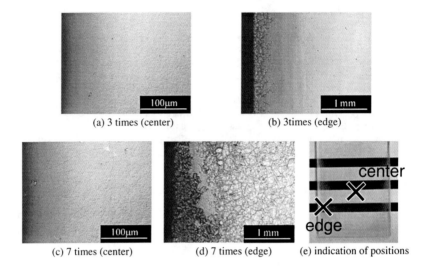

(a) 3 times (center)   (b) 3times (edge)

(c) 7 times (center)   (d) 7 times (edge)   (e) indication of positions

Figure 14. (a)-(d) Optical photomicrographs of TiO$_2$-nanoparticle-thin films deposited in ethanol. Electrophoresis condition:0.1 mA/cm^2, 50 s. (e) positions of 'center' and 'edge' in the film.

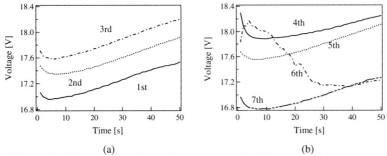

(a)   (b)

Figure 15. $V_0(t)$-t characteristics of electrophoresis using TiO₂-nanoparticle in ethanol. Electrophoresis condition:0.1 mA/cm², 50 s, (a) 1~3 times repeating, (b) 4~7 times repeating.

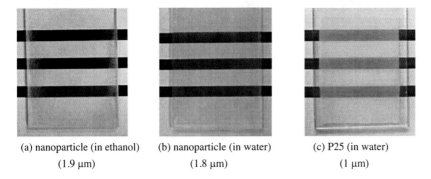

(a) nanoparticle (in ethanol)   (b) nanoparticle (in water)   (c) P25 (in water)

(1.9 μm)   (1.8 μm)   (1 μm)

Figure 16. Photographs of TiO₂-nanoparticle-thin film: Electrophoresis condition: (a) 0.1 mA/cm², 50 s repeated 7 (total electrophoresis time of 350 s) times in ethanol using nanoparticles, (b) 0.1 mA/cm², 260 s in water using nanoparticles, (c) 0.1 mA/cm², 40 s in water using P25.

found from figure 16 that the thin film with high transparency could be deposited by repeating the deposition using TiO₂-nanoparticles in ethanol with sufficient drying time. Furthermore, in order to improve the quality and to increase thickness of TiO₂-thin film, the deposition in the colloidal solutions of various viscosities controlled by changing the colloidal solution temperature is in the way of investigation.

CONCLUSION

(1) It was found from TEM image of TiO₂-nanoparticles that TiO₂-nanoparticles were connected with each other like a chain and oriented attachment mechanism was also observed. It was also found from X-ray diffraction patterns that the crystal structure of TiO₂-nanoparticles was anatase crystal structure.

(2) TiO_2-thin film with high optical transparency could be deposited using TiO_2-nanoparticles and electrophoresis deposition method.

(3) The generation of cracks correlated with the incline of $V_0(t)$-t characteristics and the thin film deposited in the small incline range of $V_0(t)$-t characteristics had cracks because mobile species which don't contribute to the transport of TiO_2-nanoparticles.

(4) The cracks were reduced using ethanol as a solvent compared to using water.

(5) It is found from the $V_0(t)$-t characteristics that TiO_2-nanoparticles-thin film with high optical transparency and no apparent cracks deposited when the electrophoresis deposition stopped in a short time. When the short time deposition was repeated in ethanol, the thickness of TiO_2-nanoparticles-thin film increased and this the film had high optical transparency and no apparent cracks.

REFERENCES

[1]T. Endo, M. Yuichi, and M. Isai, *IEICE Technical Report*, ED2008-6, CPM2008-14 SDM-26, 23-28 (2008)

[2]B. O´ Regan and M. Grätzel, *Nature*, **353**, 737-740 (1991)

[3]K. Yamamoto, M. Watanabe, and S. Yoshikado, *Key Engineering Materials,* **388**, 115-118 (2009)

[4]M. Adachi, Y. Murata, J. Takao, J. Jiu, M. Sakamoto, and F. Wang, *J. Am. Chem. Soc,* **126**, 14943-14943 (2004)

[5]K. Ito, Y. Sato, M. Adachi, and S. Yoshikado, *Key Engineering Materials* (2010), in press.

EFFECT OF NANO-SILICA ON ACID RESISTANCE PROPERTIES OF ENAMEL AND ITS CONNECTION TO ENERGY SAVING

Majid Jafari* , Javad Sarraf
Islamic Azad University, Najafabad Branch
Esfahan Iran

ABSTRACT

In oil refineries the consumption of fuel in the crude oil heating furnace is very high. Oil, gas and petrochemical industries in Iran investing now and in future to improve the energy efficiency by reduction in fuel consumption. One of the ways is using glass coated air preheating units. It is possible to reduce fuel consumption up to 15% by incorporating air preheated to the burners. Since air preheating units are installed through exhaust flue gas the chemical corrosion of metal parts is very important issue. Chemical corrosions occur by reaction of different gaseous materials such as SO_3 with H_2O in exhaust gas and formation of sulphuric acid. A formation of sulphuric acid and reaction with glass coated metal plate causes to reduce the life time of preheating unit.

In this experimental study we have tried to use amorphous nano silica particles to investigate durability of enamel against chemical attacks. Different proportions of amorphous nano silica powder were added to the commercial enamel that already has been used for coating on steel plates. After applying, sintering of enamels at different temperature was investigated and the demand properties evaluated. The results were shown that incorporating of amorphous nano-silica can improve the acid resistance of enamel without any problem on sintering of added nano silica. Nano silica dissolves in enamel at firing temperature and form uniform component in the composition of glassy layer. In this study physical, chemical and sintering of nano silica enamel were investigated.

INTRODUCTION

Most of the refinery furnaces run by different kinds of fuel, natural gas, oil and mixture of gas, and refinery waste fuels, in those cases one things is similar which is emission of CO, NO, NO_2 and sulfur[2]. Discharge fuel gases containing corrosive sulfur dioxide / trioxide gases, some parts of these gases react with atmospheric moisture and produce corrosive condensate sulfuric acid [1]. Further, pollutants in refinery furnace exhaust gases have a great variety from point of chemical components. Control of green house gas emission (GHG) is the major issue and can be calculated by following equation [3]:

$$\text{Green House Gas (tones)} = CO_{2\,tones} + (21 \times CH_{4\,tones}) + (310 \times N_2O_{tonnes}) + (1,300 \times HFC_{tones}) + (23900 \times SF_{6\,tones}) + (6500 \times PFC_{tones})$$

There are very similar situation between emission from refinery furnaces and automotive. In Table I emitted gaseous component from automotive are listed [4], and it is clear that the emission from refinery furnaces are more aggressive in pollution.

*m_jafari@iaun.ac.ir

Table I- Pollutants emitted by automotive.

Primery		
C components:	With O:	Alcoholes, aldehydes, Ketones, carboxylic acids, CO,
N compounds:	Without O:	CO_2
	Oxidized:	Alifatic, aromatic mono and polycyclic hydrocarbons
	Reduced:	NO, HNO_3
S compounds:	Oxidized:	HCN, NH_3, organic amines
		$SO_2\, H_2SO_4, SO_4$ particulate
	Reduced:	H_2S, CS_2 , organic sulphides
Other compounds	Pb (mineral or organic), asbestos , other metals (Cu, Zn)	
Secondary:	NO_2, aldehydes, O_3, PAN	

There are some ways to reduce pollutant gases by reduction in fuel consumption, the most potent way to improve efficiency and productivity is to preheat the combustion air going to the burners. The source of this heat energy is the exhaust gas stream. A heat exchanger, placed in the exhaust stack or ductwork, can extract a large portion of thermal energy in the flue gases and transfer it to the incoming combustion air. According to energy efficiency and renewable energy U.S. department of energy, fuel saving for different process temperatures evaluated as it shows in Table II [5].

Table II – Percent of fuel saving gained from using preheated combustion air.

Furnace Exhaust Temperature, °C	Preheated Air Temperature, °C					
	315	426	537	648	760	870
537	13	18	---	---	---	---
648	14	19	23	---	---	---
760	15	20	24	28	---	---
870	17	22	26	30	34	---
982	18	24	28	33	37	40
1093	20	26	31	35	39	43

The beneficent to reduce fuel consumption is huge and according to this report, it is possible to reduce fuel consumption or pollutant gas emission minimum 13% [6].

In Iran from nine refineries only two of them have air preheating units, the problem of these units are maintenance and protection from chemical corrosion. Air preheating system has recuperators with gas to gas heat exchangers installed on the stuck of the furnace. Internal plates of heat exchanger transfer heat from the outgoing gas to the incoming combustion air while it prevents mixing of two streams. This unit has three passes as it can be seen in Figure 1. The most corrosion chemical attack happen in pass1 because of reaction of sulfur gases with water at dew point temperature, since the temperature in this area is suitable to condense liquid sulfuric acid. Normally the internal plates of pass1 are coated with acid resistant enamel but since the stream of highly corroded gas during 24 hr and seven days a week is very high therefore the life time of this section is reducing to about 8 month. Despite the saving that we can obtain from air preheating heat exchanger the cost of maintenance and downtime of this system are grate concerned.

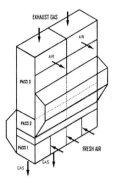

Figure 1 – Air preheating heat exchange unit with three sections, pass1, 2 and 3. The Pass1 is protected by acid resistant glass coated layer.

The materials for making air preheating unit according to our research are low carbon mild steel coated with acid resistant enamel and Corten steel. In order to increase acid resistance of internal plates of air preheating system we have tried to use amorphous nano-silica. In this study preparation, evaluation and characterization of ceramic glass coating which applied on mild steel are investigated. The coated plates were examined and characterized with optical microscope, impact test and also the important matter, acid resistance test were carried out according to ASTM and DIN standards.

EXPERIMENTAL

The acid resistance of enamels can be varying widely depending on composition and to some extent the processing used. Special formulations can provide enamels with excellent corrosion protection against aqueous solutions of most acids except hydrofluoric. Highest degree of acid resistance is obtained by sacrificing other desirable properties such as alkali resistance. The raw materials for this study were choosing from international manufacturer. Acid resistance enamel or frit supplied from commercial producer, Keskin Kimia Co. from Turkey, nano-silica from Degussa Co., Aerosil 200 grade and low carbon mild steel from Mobareke steel Co., Iran. The specification of enamel from Keskin Kimia Co is confidential but following our research it is based on SiO_2, B_2O_3, BaO, NiO, CoO and alkaline oxides. The specifications of nano-silica are shown in Table III.

Figure III. Physical and chemical properties of nano- silica [7].

Properties	Unit	Typical Value
Specific surface area (BET)	m^2/g	200 ± 25
Average primary particle size	nm	12
Tapped density	g/l	approx. 50
Bulk density	g/l	approx. 30
Moisture	wt. %	≤ 1.5
pH		≤ 1.0
SiO_2	wt. %	≥ 99.8

Different proportion of amorphous nano-silica and acid resistance enamel were prepared as it shows in Table V.

Table V- Different proportion of enamel and nano-silica.

Sample number	Enamel wt. %	Nano-silica wt. %
A1	100	---
A2	99.5	0.5
A3	95	3
A4	90	5
A5	85	10

The 300gr batch samples were prepared by accurate weighing and mixing the solid materials with water in 500 liter fast mill. To see the exact effect of nano-silica on physical and chemical properties of enamel no additional additives were used. Rheological properties of all slurries from points of milling time and viscosity were controlled and coating on plates applied by dipping and spraying techniques. The mild steel plates with diameter of 80mm and thickness of 1.5mm were used. Steel plates were prepared according to conventional method including washing to remove oil and rusted materials [8].

Green coated enamel with thickness of 0.25-0.35mm, fired over wide range of temperature (780 – 860 °C), to select good surface finish and without any surface defects. Firing and sintering at different temperature and same soaking time under or over firing condition were investigated. The thickness of the glass coat layer after firing measured by Elcometer and it was 0.2 – 0.3 mm.

Acid resistance tests were carried out according to ASTM C283, with citric acid, but since this acid for our purpose were very weak we used sulfuric acid with concentration of 10%. The samples were immersed in acid solution and test carried out in nearly boiling temperature at 95 °C for 36 hours.

RESULTS AND DISCUSSION

The samples were evaluated after firing at different temperature by their appearance since the surface color and defects shows under or over firing conditions. The maximum amount of nano-silica that could be added to compositions depended on sample preparation and firing conditions. Both enamel and nano-silica have amorphous structure therefore it was not possible to investigate residual or undissolved nano-silica in enamel by x-ray method. The best results were obtained at firing temperature of 845°C for soaking time of 5 minutes. The acid resistance results are shown in Table VI.

Table VI- Test results of coated and fired steel plates.

Composition No.	Acid resistance (wt loss in mg/cm^2)	Surface appearance	Firing temperature (°C)	Adherence
A1	3.2	Good	845	week
A2	2.5	Good	845	Good
A3	2.2	Very Good	845	Very Good
A4	2.25	Very Good	845	Very Good
A5	0.3	Good	845	Very Good

The benefits of adding nano- silica to enamel can be summarized as follow:
-   Reaction between Fe on the surface of steel and nano-silica fabricated Fe/SiO$_2$ phase [8]. The X-ray pattern of this reaction is shown in Figure 2. This reaction helps to increase strength and adhesion in the interface of metal-glass and preventing gas formation during firing. To see this effect the adhesion impact test was carried out according to DIN51155. The results were shown

that by increasing the nano silica from 0.5% to 3%, strength of the glassy layer increased effectively (Figure 3).

- As it shows in Figure 5 by introducing nano silica particles the size and number of porosity declined. This is demonstrated in optical microscope photos that obtained from cross section of fired samples.
- By incorporating nano-silica particles, packing and compaction of green enamel can be increased and it causes to increase densification during drying and sintering of enamel.
- The viscosity of glass during firing by introducing nano SiO_2 also can be increased and it helps to prevent formation and movement of gas bubbles through the glassy layer to the surface.
- SiO_2 inherently acid resistance substrate and by introducing nano particles of silica to enamel it is possible to improve this property.
- To some points dissolution of nano-silica by having very fine particle size, average 12nm, during firing in glassy media can be applicable.
- Increasing the hardness of glassy surface as it measured according to ASTM E448-82. To measure hardness of enamel four points on the surface of each sample ware tested randomly. The results are drawn in Figure 4. As it can be seen by increasing nano-silica from 0.5 to 10%, there is 30% increscent in hardness. This can be explained both, in addition of SiO_2 particles and reduction in porosity in glassy layer.

There are also some disadvantages, such as mismatching of thermal expansion of glassy surface and steel since SiO_2 reduces the thermal expansion.

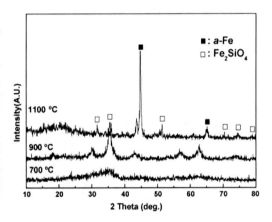

Figure 2- XRD patterns of the as-prepared nano powders of Fe and SiO_2 [9].

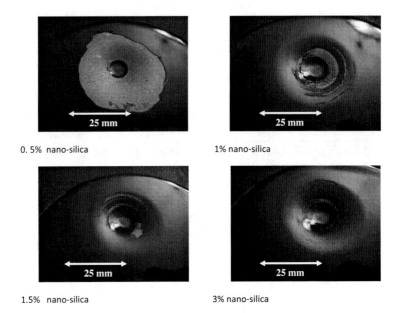

0. 5% nano-silica   1% nano-silica

1.5% nano-silica   3% nano-silica

Figure 3- The effect of nano silica increment on adhesion of glassy layer to metal.

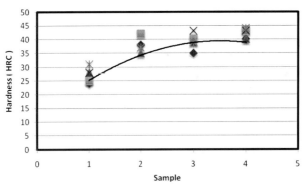

Figure 4- Increase in hardness with addition of nano-silica particles.

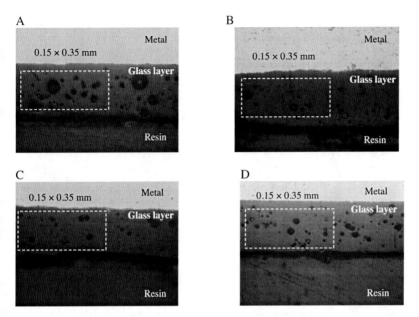

Figure 5- Polished sections of glass coated samples with different proportion of nano-silica, A: 0 %, B: 0.5 %, C: 5%, D: 10%.

CONCLUSION

The special novel idea for introducing nano-silica particles to acid resistance enamel develops resistant to 10% sulfuric acid solution. The life time of glass coated plates, has been estimated by adopting the constructed mathematical model by using Brandon's equation [10]. Variation factors in this equation are estimated, coat thickness and concentration of surface defects. The life time of a coating can be obtained by:

$$\tau_e = 0.9h \, / \, k \, (1 - 0.44 \, log \, P)$$

Where τ_e is the life time in year, h, the thickness in mm, k, the corrosion rate in mm/year, P the number of pores and large bubbles per unit area (cm^2) in the glassy coated. In experimental work if we consider the values of h and k are to be constant, according to optical photos we can measure the number and size of the porosity. Therefore the proportion of life time in unit area of 0.15 by 0.35 mm (dash line area in photos A and D) can be calculated. In Figure 5 by comparing the size and number of porosity in the samples with zero to 10% nano silica, the proportion of life time can be calculated as follow:

$$\tau_e (\, 0\% \, nano) \, / \, \tau_e(5\% \, nano) = 0.63 \, / \, 0.36 = 1.75$$

According to this calculation the life time of glassy layer can be increased by reduction in porosity as much as 1.75 times. Therefore by introducing nano-silica to the composition of enamel, we can obtain the following results.

Decline in the rate of acidic corrosion and increase in hardness in coated glassy layer have been improved. We can conclude that the nano silica particles have marked effect on improving physical and chemical properties of glassy layer and therefore life time of air preheater units in the refineries.

ACKNOWLEDGEMENTS

The author would like to thankful Mr. Khatibi and Miss Nosratollahi for various cooperation in Loab Tehran Co. Also the laboratory staff of Islamic Azad University, Najafabad branch.

REFERENCES

[1]-Keraaijveld and Gazo etal Ceram. Eng. Sci. Proc. 9 471 Phipps K J 1985 Vit Enameller 361 1988.
[2]-Sorenson S. C. "Prediction of exhaust emission from prime movers and small heating plant furnaces", National Technical Information Service, US department of commerce Springfield VA 22131, 1972.
[3]-Shell international B.V., The Hague, Netherlands "Health, safety and environment", Feb, 2007.
[4]- Azpiazu M. N., et al, "Effect of air preheating on automotive emission", Departmento de Ingenieria Energetica, E.T.S. Ingenieros. Alda. De Urquijo s/n. 48013 Bilbao, Spain 1990
[5]- Office of industrial technologies energy efficiency and renewable energy U.S. department of energy Washington, "Energy tips" , 2002.
[6]- Jae-Wook Lee[b], Byung-Yeon Park[b] and Chul-Jin Choi, "Characteristics of Fe/SiO$_2$ nanocomposite powders by the chemical vapor condensation process", University of Ulsan, P.O. Box 18, Ulsan 680-749, 2007.
[7]. www.arosil.com
[8]- John B Wachtman, "Ceramic Film and Coating," Noyes Publications, 1993.
[9]- Lee Burtand "Chemical processing of Ceramics" , Taylor and Frances Pub. 2005.
[10]. Brandon D. "Enamels glass and porcelain instrument for testing acid and neutral liquids and their vapour", ISA Journal 6, 1959.

IMMOBILIZATION OF MYOGLOBIN WITH REGENERATED SILK FIBROIN/MWCNTS ON SCREEN-PRINTED ELECTRODE: DIRECT ELECTROCHEMISTRY AND ELECTROCATALYSIS OF H2O2

Lei Zhang, Lei Shi, Wei Song, Yi-Tao Long*
School of Chemistry & Molecular Engineering, East China University of Science & Technology, Shanghai, 200237, P.R. CHINA

Electrochemical redoxprotein-based sensors represent a promising way to prepare the third generation biosensors. It is difficult for redoxproteins to exchange electrons with electrode surfaces directly and keep its biological activity in the same time. An approach to realize direct electrochemistry of proteins and enzymes is to incorporate them into films to fabricate a modified electrode surface. The thin films may provide a well-defined microenvironment for redoxproteins, and enhance the direct electron transfer between redoxproteins and electrodes.

Electrochemical methods have already proved to be powerful tools in automation and with high sensitivity, low cost and relatively short analysis time when compared with other techniques.[1, 2]. The screen-printing technology, such as Screen-Printed Carbon Electrode (SPCE), has been widely used in electrochemical research. Considering the improved repeatability of SPCE, we can expect a broaden application in electrochemistry. In the previously studies, Zen et al. have reported the exclusive applications of disposable preanodized screen-printed carbon electrode (pSPCE), which highly improved the electrochemical activity of the SPCE[3]. The pSPCE with the introduction of edge plane carbonyl groups was found to act more or less similar as an edge plane graphite or Carbon nanotubes (CNTs) electrode.

Here, we reported a novel disposable H2O2 biosensors based on the preanodized carbon electrode surface which was coated with myoglobin-silk fibroin/mutil-walled carbon nanotubes (Mb-SF/MWCNTs) compound film. Direct electron transfer between Mb and pSPCE and the electrocatalytic reduction of H2O2 was studied in detail. The immobilized Mb retained its biological activity and showed a good electrocatalytic response to the reduction of hydrogen peroxide.

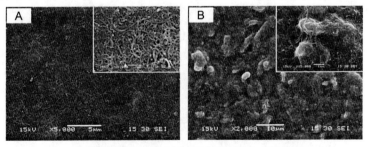

Figure 1. SEM images of SF/MWCNTs modified pSPCE(A) and
Mb-SF/MWCNTs modified pSPCE(B)

Fig. 1 shows SEM images of the SF/MWCNTs modified pSPCE surface (A) and the Mb-SF/MWCNTs modified pSPCE surface (B). It can be clearly observed that Mb was immobilized into the SF/MWCNTs matrix on the interface of pSPCE.

The effects of pH on the potential and the current of the pSPCE were studied in different phosphate buffer solution (PBS) (from pH 4.0 to 10.0). The oxidation current reached the top at pH 6.0. The slope of the linear regression equation for the potential vs. pH was gained 52.7 mV/pH, which indicated the proton and electron radio in electrode surface reaction process is 1. Fig. 2A shows cyclic voltammograms (CVs) of the Mb-SF/MWCNTs modified pSPCE in pH 6.0 PBS with the different concentration H_2O_2. A pair of well defined and nearly symmetrical redox peaks was obtained (Fig. 2 A-a), which suggested that the redox peaks were ascribed to the electrochemical reaction of Mb immobilized in the SF/MWCNTs matrix.

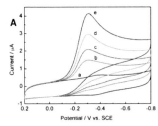 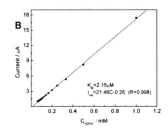

Figure 2. CVs of Mb-SF/MWCNTs modified SPCE electrocatalysis to different concentration H_2O_2 (A) (a to e: 0, 0.3, 0.5, 0.8, 1.0 mmol/L) and the fit linear of response current to concentration of H_2O_2(B)

Fig. 2B shows the amperometric response for the reduction of hydrogen peroxide on Mb-SF/MWCNTs modified pSPCE at the different concentration range. The electrocatalytic current (i_{cat}) increased linearly with increasing the concentration of H_2O_2 and reached a maximum, which suggested that the immobilized redoxprotein Mb responds to the substrate H_2O_2 in a Michaelis–Menten model. The linear range is from 1 to 149 μmol/L with a detection limit of 0.36 μmol/L. The K_M of Mb immobilized into SF/MWCNTs was estimated to be 2.15 μmol/L. Based on the experiment results, the MWCNTs enhanced greatly the electron transfer between Mb and surface of pSPCE, and the SF retained Mb biological activity for a few weeks.

In conclusions, a novel H_2O_2 biosensor is fabricated to perform a direct electrochemistry between Mb and the modified electrode. The Mb immobilized in SF/MWCNTs matrix can retain its catalytic activity toward H_2O_2. The biosensor exhibits a rapid response time, a high sensitivity and a good linear range with a low detection limit of 0.36 μmol/L. These results indicated the SF/MWCNTs matrix can provide a favorable microenvironment for direct electron transfer between the Mb and the interface of pSPCE.

ACKNOWLEDGMENT
    We greatly appreciate the support of National High Technology Research and Development Program of China (863 Project: 2008AAO6A406).

REFERENCES

[1]   C. E. Banks R. G. Compton, "New electrodes for old: from carbon nanotubes to edge plane pyrolytic graphite". *The Analyst*, vol. 131, pp. 15-21, 2006.

[2]   F. Kurusu, H. Tsunoda, A. Saito, A. Tomita, A. Kadota, N. Kayahara, I. Karube, M. Gotoh, "The advantage of using carbon nanotubes compared with edge plane pyrolytic graphite as an electrode material for oxidase-based biosensors". *The Analyst*, vol. 131, pp. 1292-1298, 2006.

[3]   K. Sudhakara Prasad, G. Muthuraman, J.M. Zen, "The role of oxygen functionalities and edge plane sites on screen-printed carbon electrodes for simultaneous determination of dopamine, uric acid and ascorbic acid". *Electrochemistry Communications*, vol. 10, pp. 559-563, 2008.

LIQUID PHASE PATTERNING AND MORPHOLOGY CONTROL OF METAL OXIDES

Yoshitake Masuda
National Institute of Advanced Industrial Science and Technology (AIST), 2266-98 Anagahora,
Shimoshidami, Moriyama-ku, Nagoya 463-8560, Japan
masuda-y@aist.go.jp   http://staff.aist.go.jp/masuda-y/index.html

ABSTRACT
    Liquid phase patterning of europium doped yttrium oxide thin films was realized. Europium doped yttrium oxide thin film was deposited on amino groups regions of a patterned self-assembled monolayer(SAM) in an aqueous solution. The thin films showed red emission (611 nm) due to photoluminescence excited by 266 nm after annealing. The micropatterning of visible-light-emitting yttrium oxide was successfully realized in an aqueous solution at environmentally friendly condition.

INTRODUCTION
    Red emitting europium-activated yttrium oxide (Y_2O_3:Eu) has been attracted much attention as both one of the most promising oxide-based red phosphor systems[1,2] due to its excellent luminescence efficiency, color purity, and stability[3] and as a model system for the study of the effect of the interplay of radiative and nonradiative processes on phosphor efficiency.
    Cubic Y_2O_3 is a good host material for rare earth ions and Y_2O_3:Eu is a superior red phosphor with a quantum efficiency of nearly 100%[2]. Yttrium oxide doped with Eu exhibits strong UV and cathode-ray-excited luminesence, which is useful in lighting and cathode ray tube, display material, tricolor fluorescent lamps[2,4,5], projection televisions[2,6], field emission displays[7,8] and laser devices, etc.[9-11]. As an oxide, it is more stable than sulfur-containing phosphors which undergo changes in their surface chemistry when interacting with the electron beam, seriously degrading their cathodoluminescent brightness and releasing gases that can poison the field emitting tips[7,12].
    In the preparation of Y_2O_3:Eu nanoparticles, many different techniques have been reported such as spray pyrolysis[11,13], chemical vapor deposition[14], sol–gel[9,15] and so on, and photoluminescence (PL) characteristics of Y_2O_3: Eu polycrystal, thin film and nanoparticle phosphors have been reported for various optical applications.
    The luminescence properties are deeply dependent on crystalline size, crystallinity, Eu concentration, uniformity of Eu distribution, etc. and the improvement and optimization of them are necessary to enhance the luminescence properties.
    We have tried to synthesize visible-light emitting Y_2O_3:Eu using solution system which has advantages in control of grain size, crystalline size and morphology, and high uniformity of Eu distribution.
    Phosphor particles (3–10 μm) without regular shape are prepared commercially by high-temperature solid-state reactions, resulting in large particles and agglomerates. The size of them is reduced by mechanically grinding/milling, which leads to the formation of nonradiative defects and the introduction of nonradiative impurities, significantly reducing luminescence efficiency. On the other hand, phosphor particles can be prepared by the solution process without the formation of nonradiative defects and the introduction of nonradiative impurities caused by grinding process. Additionally, a high degree of homogeneity is achievable in a solution, since all of the starting materials are mixed at the molecular level. Doping of Eu activator through solutions is straightforward, easy, and effective[9].
    In addition, we prepared Y_2O_3:Eu particulate films on substrates directly from the solution without the pre-preparation of Y_2O_3:Eu powders to avoid the damage of phosphors and the degradation of luminescence

properties during the particle size reduction process and the film formation process from the powders. Furthermore, micro fabrication such as two dimensional (2D) micro patterning of Y_2O_3:Eu film is indispensable to fabricate display devices. Micro patterning of Y_2O_3:Eu films can be prepared by etching technique of Y_2O_3:Eu, however, etching damage, (i.e., the change of chemical composition, Eu concentration, Eu condition, cystallinity, grain size and crystalline size, the boundary separation between a phosphor film and a substrate, cracks in phosphor films, clambling of the film, loss of the shape in the pattern and damage of substrates), causes serious degradation of luminescence properties even though as-prepared Y_2O_3:Eu powder has high luminescence performance. Recently, site-selective deposition of oxide films was proposed and nano/micro-fabrication such as 2D patterning of oxide films was realized utilizing self-assembled monolayers (SAMs) [16-20]. Molecular recognition, chemical reaction and various functions of organic head groups of SAMs were effectively applied for control of nucleation, growth and deposition of oxide materials based on basic scientific knowledge about deposition mechanism[21] and interface phenomena. These basic scientific knowledge were acquired from precise investigation of the chemical reactions of ions and molecules, the mechanisms for heterogeneous nucleation, homogeneous nucleation, growth and deposition of oxide (both of thin films and particles), the modification and functionality of head groups of SAMs, the interaction between head groups of SAMs and ions (and molecules) such as molecular recognition, chemical reaction, zeta potential, hydrophobic interaction etc., and the change of solution conditions with time such as concentration, pH, temperature etc. Nano/micro patterning of TiO_2 thin films having 200 nm line width, 100 nm interval and 70 nm thickness was realized with the site-selective deposition which is the finest pattern realized in the solution[16], and micro patterning of various oxides have been developed[22-26]. We have tried to apply high performance of SAM to realize site-selective deposition of Y_2O_3:Eu to fabricate visible-light emitting 2D micropatterns.

Here, we proposed and developed a novel process to realize site-selective deposition of Y_2O_3:Eu and fabricate 2D micropatterns of visible-light-emitting Y_2O_3:Eu[27]. The process allows us to avoid etching damage of the Y_2O_3:Eu which causes degradation of luminescence properties, and precise control of grain size, crystalline size, high uniformity of Eu distribution. Moreover, basic scientific knowledge included in this site-selective deposition will contribute to the development of solution chemistry for oxide materials which has been developed in decade to fabricate novel functional oxide materials. Control of nucleation, growth and deposition of oxides in the solution is the most basic and essential factor for solution chemistry for oxide materials, and molecular recognition of functional head groups of SAMs will provide a novel scientific field for SCOM (solution chemistry for oxide materials) and the controllability of the factors (nucleation, growth and deposition of oxides).

EXPERIMENTAL METHOD

SAM PREPARATION AND ITS MODIFICATION

Si substrate (p-type [100], 1 – 50 Ωcm, Newwingo Co., Ltd.) was cleaned ultrasonically in acetone, ethanol and deionized water for 5 min, respectively, in this order and was exposed to ultraviolet light and ozone gas for 10 min to remove organic contamination by using a UV/ozone cleaner (184.9 nm and 253.7 nm) (low-pressure mercury lamp 200 W, PL21-200, SEN Lights Co., 18 mW/cm^2, distance from a lamp 30 mm, 24°C, humidity 73 %, air flow 0.52 m3/min, 100 V, 320 W)[28-31]. APTS (3-Aminopropyltriethoxysilane)-SAM was prepared by immersing the Si substrate in an anhydrous toluene solution containing 1 vol% APTS for 1 h in N_2 atmosphere. The substrate was rinsed with a fresh anhydrous toluene in N_2 atmosphere. The substrate with SAM was baked at 120°C for 5 min to remove residual solvent and promote chemisorption of the SAM.

APTS-SAM was then irradiated by ultraviolet light (PL21-200) through a photomask (Test-chart-No.1-N type, Quartz substrate, 1.524 mm thickness, guarantee of line width 2 μm ± 0.5 μm, Toppan Printing Co., Ltd.) for 10 min. UV irradiation modified amino groups to silanol groups forming a pattern of amino groups regions and silanol groups regions. Patterned APTS-SAM having amino regions and silanol regions was used as a template for patterning of yttrium oxide. Initially deposited APTS-SAM showed water contact angles of 48°. UV-irradiated surfaces of SAM was, however, wetted completely (contact angle <5°). This suggests that SAM of APTS was modified to hydrophilic OH group surfaces by UV irradiation.

CHARACTERIZATION

pH and temperature of solution were measured and recorded by pH meter (Cyberscan 1100, Eutech Instruments Pte Ltd.) connected to a computer. Particles size distribution of homogeneously nucleated particles in the solution was measured by electrophoretic light scattering equipment (ELS-8000, Otsuka Electronics Co., Ltd.). After having been immersed in the solution, the substrates were rinsed with distilled water and observed by a scanning electron microscope (SEM; S-3000N, Hitachi, Ltd.), and a scanning probe microscope (SPI 3800N, Seiko Instruments Inc.) that was operated in AFM (atomic force microscopy) tap mode to observe the topography of the surface. AFM scans were operated at room temperature under ambient air. The ratio of Y : Eu was evaluated by energy dispersive X-ray analysis (EDX; EDAX Falcon, EDAX Co. Ltd.), which is built into SEM. Surface of thin films was evaluated by an X-ray photoelectron spectroscopy (XPS) (ESCALAB 210, VG Scientific Ltd.) in which the X-ray source (MgKα, 1253.6 eV) was operated at 15 kV and 18 mA, and the analysis chamber pressure was 1-3×10-7 Pa. Crystal phase of phase transition were evaluated by an X-ray diffractometer (XRD; RINT-2100, Rigaku) with CuKα radiation (40 kV, 30 mA) and Ni filter plus a graphite monochromator. Photoluminescence images of the films were taken with a digital camera (Coolpix 8400, 8.0 megapixels, Nikon Corporation) and photoluminescence spectra were evaluated by a fluorescence spectrometer (F-4500, excitation wavelength 350 nm, Xe lamp, Hitachi, Ltd.).

RESULTS AND DISCUSSION

SYNTHESIS OF YTTRIUM OXIDE

The patterned APTS-SAM was immersed in an aqueous solution containing Y(NO$_3$)$_3$ · 6H$_2$O (4 mM), Eu(NO$_3$)$_3$ · 6H$_2$O (0.4 mM) and NH$_2$CONH$_2$(50 mM) at 25 °C. The solution was heated to 77 °C gradually as shown in Fig. 1 since urea (NH$_2$CONH$_2$) decomposed to form ammonium ions (NH$_4^+$) above 70 °C (Eq. (a)). The decomposition of urea at elevated temperature plays an essential role in the deposition of yttrium oxide. The aqueous solution of urea yields ammonium ions and cyanate ions (OCN$^-$) at temperature above 70 °C[32].

$$NH_2\text{-}CO\text{-}NH_2 \rightleftarrows NH_4^+ + OCN^- \qquad \text{(a)}$$

Cyanate ions reacts rapidly according to Eq. (b).

$$OCN^- + 2H^+ + H_2O \longrightarrow CO_2 + NH_4^+ \qquad \text{(b)}$$

Yttrium ions are weakly hydrolyzed[33,34] in water to $YOH(H_2O)_n^{2+}$.

$$Y(NO_3)_3 \cdot 6H_2O \longrightarrow [YOH(H_2O)_n]^{2+} + 3NO_3^- + H^+ + (5-n)H_2O \qquad (c)$$

The resulting release of proton (H^+) and/or hydronium ions (H_3O^+) accelerates urea decomposition (Eq. (b)). The precipitation of the amorphous basic yttrium carbonate ($Y(OH)CO_3 \cdot xH_2O$, x=1) can be described by the following reaction[35,36].

$$[YOH(H_2O)_n]^{2+} + CO_2 + H_2O \rightleftarrows Y(OH)CO_3 \cdot H_2O + 2H^+ + (n-1)H_2O \qquad (d)$$

The controlled release of carbonate ions by urea decomposition causes deposition of basic yttrium carbonate once the critical supersaturation, in terms of reacting component is achieved. Since the decomposition of urea is quite slow, the amount needed to reach supersaturation within a given period of time must be considerably higher than the stoichiometric amount of yttrium ions, as revealed by previous studies of lanthanide compounds[37].

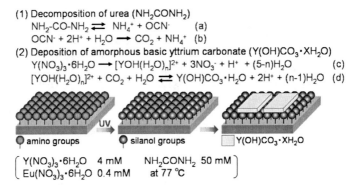

(1) Decomposition of urea (NH_2CONH_2)
$$NH_2\text{-}CO\text{-}NH_2 \rightleftarrows NH_4^+ + OCN^- \qquad (a)$$
$$OCN^- + 2H^+ + H_2O \longrightarrow CO_2 + NH_4^+ \qquad (b)$$
(2) Deposition of amorphous basic yttrium carbonate ($Y(OH)CO_3 \cdot XH_2O$)
$$Y(NO_3)_3 \cdot 6H_2O \longrightarrow [YOH(H_2O)_n]^{2+} + 3NO_3^- + H^+ + (5-n)H_2O \qquad (c)$$
$$[YOH(H_2O)_n]^{2+} + CO_2 + H_2O \rightleftarrows Y(OH)CO_3 \cdot H_2O + 2H^+ + (n-1)H_2O \qquad (d)$$

amino groups   silanol groups $Y(OH)CO_3 \cdot XH_2O$

$$\begin{bmatrix} Y(NO_3)_3 \cdot 6H_2O & 4\ mM & NH_2CONH_2 & 50\ mM \\ Eu(NO_3)_3 \cdot 6H_2O & 0.4\ mM & at\ 77\ ^\circ C \end{bmatrix}$$

Figure 1 Conceptual process for site-selective deposition of visible-light emitting Y_2O_3:Eu thin films using a self-assembled monolayer.

Temperature of the solution increased gradually and reached to 77 °C for about 80 min as shown in Fig. 2. The solution was kept at about 77 °C during deposition period. pH of the solution increased from pH 5.2 to pH 5.8 for about 90 min and gradually decreased to pH 5.6. Temperature and pH increased for initial 90 min and are stable after 90 min. Average particle size homogeneously nucleated in the solution at 100 min was about 226.5 nm and increased to 261.8 nm at 150 min, 282.3 nm at 180 min, 310.3 nm at 210 min, 323.4 nm at 240 min (Fig. 3). Particles would nucleate and grow after the solution temperature reached to above 70 °C because urea decomposes above 70 °C to form carbonate ions[32] which causes deposition of basic yttrium carbonate[33-36]. They would grow rapidly at the beginning of growth period and decrease its growth rate exponentially (Fig. 3). The decrease of growth rate would be caused by the decrease of super saturation degree influenced by decrease of solution concentration.

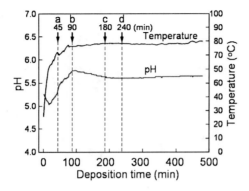

Figure 2 Time variation of pH and temperature of the solution.

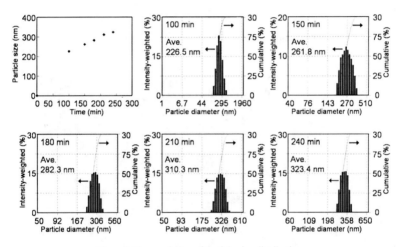

Figure 3 Time variation of particle size distribution.

PATTERNING OF YTTRIUM OXIDE

The oxide thin films were observed on amino regions of a patterned SAM after the immersion in an aqueous solution (Fig. 4). Depositions showed white contrast, on the other hand, silanol regions without deposition showed black contrast in SEM observation. Narrow lines of depositions having 10-50 μm width were successfully fabricated in an aqueous solution. Patterned APTS-SAM showed high ability for site-selective deposition of yttrium oxide in solution systems.

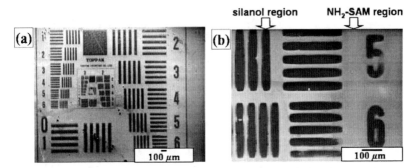

Figure 4 SEM micrographs of patterned Y_2O_3:Eu thin films

Yttrium, oxygen and carbon were observed from as-deposited thin films on amino regions, and silicon was detected from non-covered silanol regions (Fig. 5). The as-deposited thin films would be composed from yttrium, oxygen and carbon, and silicon and small amount of oxygen were detected from silicon wafer covered with natural oxide layer (amorphous SiO_2). Clear mapping image of europium wasn't obtained due to its small concentration in the film.

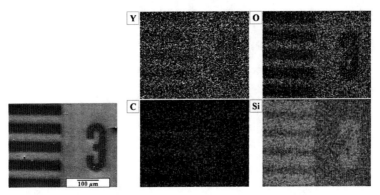

Figure 5 Elemental analysis of patterned Y_2O_3:Eu thin films using EDX: SEM image and characteristic X-ray images of Y, O, C and Si.

Chemical composition was further evaluated quantitatively by EDX analysis (Fig. 6). Yttrium, europium, oxygen, carbon and silicon were observed from depositions. Silicon and oxygen (mole ratio Si : O = 1 : 2) were detected from Si substrate covered with natural oxide layer (amorphous SiO_2) and the deposition would be composed of yttrium, europium, oxygen and carbon. Molecular ratio of yttrium to europium was determined to be 100 : 8. The content of europium was in the range of our expected. Y_2O_3:Eu with atomic ratio Y : Eu = 100 : ~ 8 were reported to have strong photoluminescence[38,39]. Carbon would be detected from residual urea and by-product containing carbon such as carbonates.

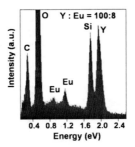

Figure 6 Elemental analysis of Y$_2$O$_3$:Eu thin films deposited for 90 min.

Amino regions covered with depositions showed thin films composed from many large particles (about 100-300 nm in diameter) and very high roughness (RMS 25.6 nm) (Fig. 7). Silanol regions, on the other hand, showed only nano-sized small particles (about 10-50 nm in diameter) and very low roughness (RMS 1.7 nm). High site-selectivity of deposition and the big difference in surface morphology and roughness were clearly shown by AFM observation.

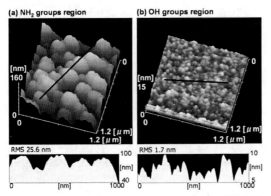

Figure 7 AFM images of (a) Y$_2$O$_3$:Eu thin films on NH$_2$ groups regions and (b) OH groups regions without deposition.

Yttrium wasn't detected by XPS from the substrate immersed for 45 min (Fig. 8-1), however, it was clearly observed from that immersed for 90 min (Fig. 8-2a). This showed the deposition was begum after 45 min and before 90 min. Solution temperature was shown to reach to 70 °C at around 45 min (Fig. 2) and would begin to decompose to release carbonate ions which causes deposition of basic yttrium carbonate. The deposition mechanism evaluated by XPS is consistent with the change of solution temperature, decomposition temperature of urea and chemical reaction of this system. Binding energy of Y 3d5/2 spectrum from the deposition (158.2 eV) (Fig. 8-2a) was higher than that of metal yttrium (155.8eV)[40]. The spectrum shifted to lower binding energy (156.7 eV) after annealing at 800 °C in air for 1 h (Fig. 8-2b) and is similar to that of Y$_2$O$_3$ (157.0eV)[41]. The binding energies of Y 3d5/2 spectra in as-deposited films and annealed films were higher than that of metal yttrium possibly due to the chemical bonds between yttrium ions and oxygen ions. The chemical shift of Y 3d5/2 binding energy by annealing is consistent with crystallization of as-deposited films to crystalline Y_2O_3. As-deposited films showed C1s spectra at 284.6 eV and 289.7 eV (Fig. 8-3a), and the latter disappeared by the annealing (Fig. 8-3b). The former spectrum (284.6 eV) would be surface contamination adhered at the last minute before XPS measurement because its binding energy is similar to that in common surface contamination and it was detected even after annealing. The latter spectrum (289.7 eV) would be caused from carbon ions bind to other ions in the films such as oxygen. The latter spectrum (289.7 eV) was thus disappeared by the annealing which induced crystallization of the films.

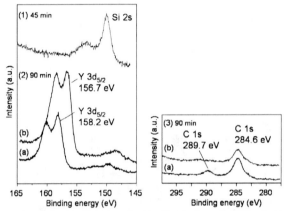

Figure 8 XPS spectra of Y_2O_3:Eu thin films deposited for (1) 45 min or (2, 3) 90 min (a) before and (b) after annealing at 800℃ for 1 h.

As-deposited film was shown to be an amorphous phase (Fig. 9a) by XRD measurement. The film showed no diffraction peak after annealing at 400 °C for 1 h, however, it showed 222, 400 and 440 diffraction peaks of crystalline cubic Y_2O_3[42] without any additional phase after annealing at 600 °C for 1 h and the intensities of diffraction peaks increased further by the annealing at 800 °C for 1 h (Fig. 9b). The film was shown to be a polycrystalline Y_2O_3 film constructed from randomly deposited Y_2O_3 particles without crystalline orientation. The crystallization by annealing confirmed from XRD measurement is consistent with XPS evaluation.

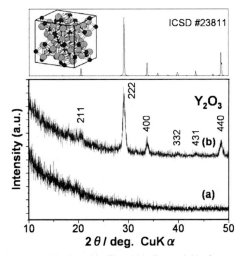

Figure 9 XRD patterns of Y_2O_3:Eu thin films thin films (a) before and (b) after annealing at 800°C for 1 h.

The Y_2O_3 films weren't removed from the silicon substrate by debonding test with scotch tape and by ultrasonication for 5 min in water to show strong adhesion between films and substrate after annealing at 800 °C for 1 h though as-deposited films were removed from the substrate on the test.

The thin film annealed at 800 °C for 1 h, i.e., crystalline Y_2O_3:Eu thin film, was shown to be excited by 230-250 nm (center: 243 nm) to emit red light photoluminescence centered at 611 nm in fluorescence excitation spectrum (Fig. 10a). The as-deposited film nor the film annealed at 400 °C for 1 h showed no photoluminescence, on the other hand, the films annealed at 600 °C or 800 °C for 1 h emitted a light centered at 617 nm by 250 nm in fluorescence emission spectra (Fig. 10b). Fluorescence intensity of the film annealed at 800 °C is stronger than that of the film annealed at 600 °C. Fluorescence intensity would increase by the crystallization and crystal growth by the heat treatments, and they are consistent with crystallization observed by XRD measurements. The spectra are described by the well-known $^5D_0$–^7F_J line emissions (J=0, 1, 2, ...) of the Eu^{3+} ion with the strongest emission for J= 2 at 612 nm. The thin film annealed at 800 °C produced visible red light photoluminescence by excitation from Nd: YAG laser (266 nm) (Fig. 10 Insertion). White square shows edges of Y_2O_3:Eu thin film and red color shows visible red emission from the irradiated area on the substrate.

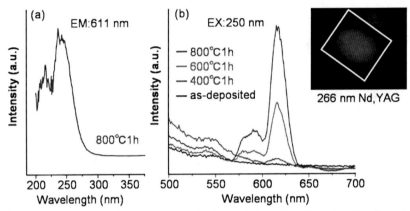

Figure 10 (a) Fluorescence excitation spectrum (emission: 611 nm) for Y₂O₃:Eu thin film after annealing at 800°C for 1 h. (b) Fluorescence emission spectra (excitation: 250 nm) for Y₂O₃:Eu thin films before and after annealing at 400, 600 or 800°C for 1 h. Insertion: Photoluminescence image for Y₂O₃:Eu thin film annealed at 800°C for 1 h (excitation: 266 nm).

CONCLUSIONS

We have proposed a novel process to fabricate visible red light emitting Eu doped Y₂O₃ and its micropattern using a self-assembled monolayer and an aqueous solution system. The patterned APTS-SAM having amino groups regions and silanol groups regions realized site-selective deposition of yttrium oxide in an aqueous solution. The deposited films were crystallized by the annealing at 600 °C or 800 °C for 1 h. Crystalline Y₂O₃:Eu produced visible red light photoluminescence centered at 611 nm by excitation from Nd: YAG laser (266 nm). This study showed high potential of aqueous solution systems and self-assembled monolayers to fabricate functional metal oxide thin films and their micropatterns.

REFERENCES

[1] S. L. Jones, D. Kumar, R. K. Singh, P. H. Holloway, and P. K. Singh, "Luminescence of pulsed laser deposited Eu doped yttrium oxide films," *Appl. Phys. Lett.*, **71** 404-06 (1997).

[2] G. Blasse and B. C. Grabmaier, "Luminescent Materials,"). Berlin: Springer. (1994).

[3] G. Wakefield, E. Holland, P. J. Dobson, and T. L. Hutchison, "Luminescence properties of nanocrystalline Y2O3 : Eu," *Adv. Mater.*, **13**[20] 1557-60 (2001).

[4] T. E. Peters, R. G. Pappalardo, and R. B. J. Hunt, "Lamp Phosphors," pp. 313. In Solid State Luminescence. Edited by A. H. Katai. Chapman & Hall, London, (1993).

[5] C. R. Ronda, "Recent achievements in research on phosphors for lamps and displays," *J. Lumin.*, **72-74** 49-54 (1997).

[6] G. Blasse, "Phosphors for other applications," pp. 349. In Solid State Luminescence. Edited by A. H. Katai. Chapman & Hall, London, (1993).

[7] P. H. Holloway, T. A. Trottier, B. Abrams, C. Kondoleon, S. L. Jones, J. S. Sebastian, W. J. Thomes, and H. Swart, "Advances in field emission displays phosphors," *J. Vac. Sci.Technol. B*, **17**[2] 758-64 (1999).

[8]E. T. Goldburt, V. A. Bolchouchine, B. N. Levonovitch, and N. P. Sochtine, "Highly efficient cathodoluminescent phosphors and screens for mid- and high-voltage field emission displays," *J. Vac. Sci. Technol. B*, **17**[2] 765-68 (1999).

[9]R. P. Rao, "Preparation and Characterization of Fine-Grain Yttrium-Based Phosphors by Sol-Gel Process," *J. Electrochem. Soc.*, **143** 189-97 (1996).

[10]S. Qiang, C. Barthou, and J. P. Denis, "Luminescence and energy transfer in Y2O3 CO-doped with Bi3+ and Eu3+," *J. Lumin.*, **28** 1-11 (1983).

[11]C. Y. Xu, B. A. Watkins, R. E. Sievers, X. P. Jing, P. Trowga, C. S. Gibbons, and A. Vecht, "Submicron-sized spherical yttrium oxide based phosphors prepared by supercritical CO2-assisted aerosolization and pyrolysis," *Appl. Phys. Lett.*, **71** 1643-45 (1997).

[12]S. Itoh, H. Toki, Y. Sato, K. Morimoto, and T. Kishino, "The ZnGa2O4 Phosphor for Low-voltage Blue Cathodoluminescence," *J. Electrochem. Soc.*, **138**[5] 1509-12 (1991).

[13]Y. C. Kang, H. S. Roh, and S. B. Park, "Morphology of oxide phosphor particles prepared by colloidal seed-assisted spray pyrolysis," *J. Electrochem. Soc.*, **147**[4] 1601-03 (2000).

[14]L. D. Sun, J. Yao, C. Liu, C. Liao, and C. H. Yan, "Rare earth activated nanosized oxide phosphors: synthesis and optical properties," *J. Lumin.*, **87-89** 447-50 (2000).

[15]M. H. Lee, S. G. Oh, and S. C. Yi, "Preparation of Eu-doped Y2O3 luminescent nanoparticles in nonionic reverse microemulsions," *J. Colloid Interface Sci.*, **226**[1] 65-70 (2000).

[16]Y. Masuda, N. Saito, R. Hoffmann, M. R. De Guire, and K. Koumoto, "Nano/micro-patterning of anatase TiO2 thin film from an aqueous solution by site-selective elimination method," *Sci. Tech. Adv. Mater.*, **4** 461-67 (2003).

[17]Y. Masuda, T. Sugiyama, H. Lin, W. S. Seo, and K. Koumoto, "Selective deposition and micropatterning of titanium dioxide thin film on self-assembled monolayers," *Thin Solid Films*, **382** 153-57 (2001).

[18]Y. Masuda, Y. Jinbo, T. Yonezawa, and K. Koumoto, "Templated Site-Selective Deposition of Titanium Dioxide on Self-Assembled Monolayers," *Chem. Mater.*, **14**[3] 1236-41 (2002).

[19]Y. Masuda, T. Sugiyama, and K. Koumoto, "Micropattening of anatase TiO2 thin films from an aqueous solution by site-selective immersion method," *J. Mater. Chem.*, **12**[9] 2643-47 (2002).

[20]Y. Masuda, S. Ieda, and K. Koumoto, "Site-Selective Deposition of Anatase TiO2 in an Aqueous Solution Using a Seed Layer," *Langmuir*, **19**[10] 4415-19 (2003).

[21]Y. Masuda, T. Sugiyama, W. S. Seo, and K. Koumoto, "Deposition Mechanism of Anatase TiO2 on Self-Assembled Monolayers from an Aqueous Solution," *Chem. Mater.*, **15**[12] 2469-76 (2003).

[22]N. Shirahata, Y. Masuda, T. Yonezawa, and K. Koumoto, *Langmuir*, **18**[26] 10379-85 (2002).

[23]N. Saito, H. Haneda, T. Sekiguchi, N. Ohashi, I. Sakaguchi, and K. Koumoto, "Low-Temperature Fabrication of Light-Emitting Zinc Oxide Micropatterns Using Self-Assembled Monolayers," *Adv. Mater.*, **14**[6] 418-21 (2002).

[24]T. Nakanishi, Y. Masuda, and K. Koumoto, "Site-Selective Deposition of Magnetite Particulate Thin Films on Patterned Self-assembled Monolayers," *Chem. Mater.*, **16** 3484-88 (2004).

[25]J. H. Xiang, Y. Masuda, and K. Koumoto, "Fabrication of Super-Site-Selective TiO2 Micropattern on a Flexible Polymer Substrate Using a Barrier-Effect Self-Assembly Process," *Adv. Mater.*, **16**[16] 1461-64 (2004).

[26]Y. F. Gao, Y. Masuda, T. Yonezawa, and K. Koumoto, "Site-Selective Deposition and Micropatterning of SrTiO3 Thin Film on Self-Assembled Monolayers by the Liquid Phase Deposition Method," *Chem. Mater.*, **14** 5006-14 (2002).

[27]Y. Masuda, M. Yamagishi, and K. Koumoto, "Site-Selective Deposition and Micropatterning of Visible-Light-Emitting Europium-Doped Yttrium Oxide Thin Film on Self-Assembled Monolayers," *Chem.*

Mater., **19** 1002-08 (2007).

[28]Y. Masuda, T. Itoh, and K. Koumoto, "Self-assembly and Micropatterning of Spherical Particle Assemblies," *Adv. Mater.*, **17**[7] 841-45 (2005).

[29]Y. Masuda, T. Itoh, and K. Koumoto, "Self-Assembly Patterning of Silica Colloidal Crystals," *Langmuir*, **21** 4478-81 (2005).

[30]Y. Masuda, T. Itoh, M. Itoh, and K. Koumoto, "Self-Assembly Patterning of Colloidal Crystals Constructed from Opal Structure or NaCl Structure," *Langmuir*, **20**[13] 5588-92 (2004).

[31]Y. Masuda, K. Tomimoto, and K. Koumoto, "Two-Dimensional Self-Assembly of Spherical Particles Using a Liquid Mold and Its Drying Process," *Langmuir*, **19**[13] 5179-83 (2003).

[32]W. H. R. Shaw and J. J. Bordeaux, "The Decomposition of Urea in Aqueous Media," *J. Am. Chem. Soc.*, **77** 4729-33 (1955).

[33]D. E. Ryabchikov and V. A. Ryabukhin, "Analytical Chemistry of Yttrium and the Lanthanide Elements," pp. 50-60). Ann Arbor, MI: Humphrey Science. (1970).

[34]C. F. Baes and R. E. Mesmer, "The Hydrolysis of Captions," pp. 129-46). New York: Wiley. (1976).

[35]B. Aiken, W. P. Hsu, and E. Matijevic, "PREPARATION AND PROPERTIES OF MONODISPERSED COLLOIDAL PARTICLES OF LANTHANIDE COMPOUNDS.3. YTTRIUM(III) AND MIXED YTTRIUM(III)-CERIUM(III) SYSTEMS," *J. Am. Ceram. Soc.*, **71**[10] 845-53 (1988).

[36]M. Agarwal, M. R. DeGuire, and A. H. Heuer, "Synthesis of yttrium oxide thin films with and without the use of organic self-assembled monolayers," *Appl. Phys Lett.*, **71**[7] 891-93 (1997).

[37]E. Matijevic and W. P. Hsu, "Preparation and Properties of Monodispersed Colloidal Particles of Lanthanide Compounds. 1. Gadolinium, Europium, Terbium, Samarium, and Cerium (III)," *J. Colloid Interface Sci.*, **118**[2] 506-23 (1987).

[38]P. K. Sharma, M. H. Jilavi, R. Nass, and H. Schmidt, "Tailoring the particle size from · -nm scale by using a surface modifier and their size effect on the fluorescence properties of europium doped yttria," *J. Lumin.*, **82** 187-93 (1999).

[39]M. G. Kwaka, J. H. Parkb, and S. H. Shon, "Synthesis and properties of luminescent Y2O3:Eu (15-25 wt%) nanocrystals," *Solid State Commun.*, **130** 199-201 (2004).

[40]J. C. Fuggle and N. Martensson, "Core-Level Binding Energies in Metals," *J. Electron Spectrosc. Relat. Phenom.*, **21** 275 (1980).

[41]C. D. Wagner, "Practical Surface Analysis," pp. 595 in 2 ed. Vol. 1): John Wiley. (1990).

[42]M. G. Paton and E. N. Maslen, "A Refinement of the Crystal Structure of Yttria, ICSD #23811," *Acta Crystallographica*, **1** 1948-23 (1967).

ROLE OF NANO-STRUCTURED DOMAIN DERIVED FROM ORGANICALLY MODIFIED SILICATE IN ELECTROCATALYSIS

P. C. Pandey, D. S. Chauhan, V. Singh
Department of Applied Chemistry, Institute of Technology, Banaras Hindu University,
Varanasi-221005, India

ABSTRACT

The roles of nanostructured network derived from organically modified silicate encapsulating the redox mediator along with manipulating the nano-structured network are reported. Oneway this generates electrocatalytic sites and the otherway contributes in electo-synthesis of polyaniline composite with variable redox properties and microstructure. The nanostuctured network of the ormosil was manipulated by introduction of a well-known redox mediator prussian blue. Polyaniline (PAni) with excellent electrochemical behavior was grown within these modified electrodes and the process of electropolymerization was found as a function of redox mediator's characteristics. The resulting polymers were analyzed by cyclic voltammetry and spectro-electrochemistry and the results again provided remarkable dependence on the property of redox mediator and palladium content justifying similar conclusion. Further, the present ormosil was also used to study the oxidation of hydrogen peroxide. The effect of electrocatalysis was also studied by introducing palladium within nano-structured network.

INTRODUCTION

Polyaniline has been widely studied because of its potential applications in electrorheological fluids[1, 2], sensors[3], electrostatic discharge[4], and anticorrosion coatings[5]. Accordingly several routes of aniline polymerization in both aqueous and non-aqueous solvents with varying media compositions are available[6-9]. The electrochemical oxidation of aniline has typically been carried out in aqueous medium galvanostatically[10, 11], potentiostatically[6, 7, 12], or through cycling of the potential (potentiodynamic) of the substrate anode between suitable potential range versus Ag/AgCl or saturated calomel electrode (SCE)[7-9, 13, 14]. We have studied in details the potentostatic and potentiodynamic mode of electropolymerization of aniline[6, 7] in which several conventional electron transfer mediators showed reversible electrochemistry. In solution, the coupling of reversible redox electrochemistry together electrochemistry of aniline polymerization is difficult to observe due to fast dynamics of polymeric intermediates and redox couple of the electron transfer relays. On the other hand if redox mediators could be confined within nano-geometry of solid-state matrix i.e., organically modified silicate (ORMOSILs) matrix within which electropolymerization of aniline is triggered, it might be feasible to record such finding if the redox mediators are stable within selected potential range.

Earlier studies on the formation of conducting polymers together sol-gel glasses have been reported by Cox et al[15, 16] and others[17-23] following three different approaches. In first one, chemically prepared polymers were dissolved and mixed with a sol that was subsequently processed into a solid material[17, 18]. Second, the silica precursor was organically modified with a monomer; subsequent to the gelation, the polymerization was performed either chemically or electrochemically[19-21]. The third general method was to form a thin film of silica on an indium tin oxide electrode, immerse the system in a solution that contains the monomer, and perform the polymerization electrochemically[22, 23].

Further, the role of nano-science in combination with the property of Prussian blue for probing hydrogen peroxide has been gaining greater attention of world scientists[24-26]. All these reports revealed the applicability of electrocatalytic layer in electro-detection of hydrogen peroxide. These contributions justify the necessity of nano-structured domain in electrocatalysis of hydrogen peroxide. We have studied in details on such systems encapsulating redox active moieties within ormosil network derived from 3-aminopropyltrimethoxysilane and 2-3-epoxycyclohexylpropyltrimethoxy silane[27-37]. Further introduction of 3-glycidoxypropyltrimethoxysilane into such matrix has an extra advantage for enhancing the material property in two different orientations; (i) structural densification, and (ii) possibility of introducing some novel electrocatalysts such as palladium chloride into nano-structured network. The latter could be availed by selective interaction of palladium chloride with the glycidoxy-residue of 3-glycidoxypropyltrimethoxysilane.

The present article justifies the contribution on these lines and indeed very interesting finding on the dependence of electrochemical polymerization versus electrochemistry of a nventional redox mediator i.e., Prussian blue. Nano-structured matrices derived through sol-gel process[34, 35] are chosen to confine the redox relays and the electropolymerization of aniline is conducted within the ormosil network. This resulted in coupling of the redox processes of the mediator alongwith electrochemistry of aniline oxidation. Additionally, an electrocatalyst palladium chloride was also introduced within the solid-state network of ormosils to justify the role of electrocatalysis on electropolymerization of aniline. The results justify novel finding on aniline electropolymerization and dependence of its properties on electrochemistry of Prussian blue and remarkable dependence on presence of palladium. The results on electrochemical characterization based on cyclic voltammetry and spectro-electrochemical measurements justified the similar conclusion. Further the ormosil was modulated in order to generate an electrocatalytic site leading to better electrochemistry in electro-analytical detection of hydrogen peroxide.

EXPERIMENTAL

Aniline, 3-Aminopropyltrimethoxysilane, 3-Glycidoxypropyltrimethoxysilane, Palladium chloride were obtained from Aldrich Chemical Co. 2-(3, 4-Epoxycyclohexyl) ethyltrimethoxysilane was obtained from Fluka. Potassium Ferricyanide, Ferrous Sulphate and Hydrogen peroxide (30%) were obtained from Merck. All other chemicals employed were of analytical grade. Aniline was distilled under vacuum prior to use. The aqueous solutions were prepared in triply-distilled water.

A typical ormosil film was prepared by adding alkoxysilane precursors, redox mediator, hydrochloric acid and distilled water in the following composition: 3-Aminopropyltrimethoxysilane = 70 µL, 2-(3, 4-Epoxycyclohexyl)ethyltrimethoxysilane = 10 µL, 0.1 M HCl = 5 µL, and distilled water = 180 µL; The mixture was vigorously stirred for 5 min. To this, a mixture of 0.05 M potassium ferricyanide and a freshly prepared solution of 0.05 M ferrous sulphate 60 µL each was added. A second ormosil film was also prepared to introduce palladium contained a solution of 3-Glycidoxypropyltrimethoxysilane and palladium chloride 10 µL each. This solution was added to above composition and water was adjusted accordingly. An aliquot of 10 µL of the suspension was coated on the ATO electrodes (resistance 200-250 Ω) with a micropipette, and the electrodes were air-dried for 8-10 hours to ensure complete hydrolysis and gelation resulting into ormosil film. All the experiments were performed at room temperature.

Polyaniline was synthesized electrochemically using single compartment cell equipped with three electrodes viz. ATO plate as working electrode, Ag/AgCl as reference electrode (Orion, Beverly, MA, USA) and Pt plate as the counter electrode. All the electrochemical work was done with an Electrochemical Workstation Model 1700C, CH Instruments Inc., TX, USA. Polyaniline was deposited potentiodynamically over ormosil modified ATO electrodes from 1 M HCl with typical concentration of 0.1 M aniline by cycling the potential between -0.2 to 1.0 V versus Ag/AgCl.

Cyclic voltammetry studies of PAni films formed electrochemically over ATO electrodes were performed by cycling the potential between -0.2 to 1.0 V versus Ag/AgCl in 1 M HCl at various scan rates viz. 0.01, 0.02, 0.05, 0.10 and 0.20 V/s. The spectro-electrochemistry experiments were performed on PAni coated ATO electrodes by cycling the potential between -0.2 to 1.0 V in 1 M HCl at 0.005 V/s simultaneously recording the electronic spectra for a single cycle using optical fiber based spectrophotometer Model DT 1000 CE UV/VIS Light Source.

RESULTS AND DISCUSSION

Role of Nano-Structured Domain during Electropolymerization of Aniline

Although there have been numerous reports available on the synthesis of PAni in aqueous media over conventional electrodes, however, the structure of PAni has been found to be dependent on the physics of the matrix within which polymerization process is conducted. The current research program concerns to justify the role of nano-structured network, during electropolymerization of aniline. In order to justify such event, it was planned to understand the process of electropolymerization on the surfaces of bare ATO electrode (Fig. 1) and ormosil-modified ATO electrode without redox mediator (Fig. 2) under similar experimental conditions. The results lead to following conclusions; (i) the process of electropolymerization is relatively much faster on bare electrode surface as compared to that on ormosil-modified electrode, (ii) the number of potentiodynamic cycles for detectable polymer properties on bare ATO is 20 whereas the same on ormosil-modified electrode is 100; (iii) The redox behavior of the polymer made on bare ATO surface is much poorer as compared that of ormosil-modified electrode. These findings suggested that incorporation of nano-structured domains during electropolymerization process generates patterned polymer structure with excellent redox behavior basically due to controlled growth of polyaniline domains within nano-structured network. This is well supported with the result of polymerization in sol-gel without mediator since the polymer in this case took one hundred cycles to exhibit good redox behavior. This is due to the generation of a patterned nano-structure network resulting in controlled growth of polyaniline. The void spaces of the sol-gel matrix acted as a template for the growing PAni and resulted in a matrix in which the polymer was more ordered than in films synthesized without any spatial restriction i.e. on bare ITO. The pores of ormosil matrix could act as an effective template that may lead to an ordered nano-structure of the produced material. The enhancements in molecular and super-molecular order in such materials could lead to electronic conductivities higher than the conventional forms (i.e., powders or thin films). In addition, conducting polymers synthesized in such manner could be used as nanoelectrodes.

Thus, to enhance the rate of polymerization of aniline and to understand the interaction between redox mediator and different chemical states of PAni[13], PAni was grown within ormosil matrix encapsulating prussian blue (Fig. 3). The voltammograms clearly demonstrate redox behavior of the mediator modulating the growth of polyaniline. In the presence of prussian blue, the number of potentiodynamic cycles required for detectable polymer properties are 35 which is much better than that recorded withour mediator.

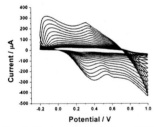

Fig. 1. Electropolymerization of 0.1 M aniline in 1 M HCl at 0.05 V/s between -0.2 V to 1.0 V versus Ag/AgCl on bare ATO electrode.

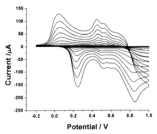

Fig. 2. Electropolymerization of 0.1 M aniline in 1 M HCl at 0.05 V/s between -0.2 V to 1.0 V versus Ag/AgCl on ATO modified with bare sol gel.

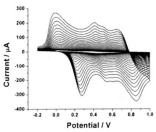

Fig. 3. Electropolymerization of 0.1 M aniline in 1 M HCl at 0.05 V/s between -0.2 V to 1.0 V on ATO modified with ormosil containing prussian blue.

Electrochemistry of PAni grown within Ormosil Matrix

The next stage of investigation is to understand the variable electrochemistry of PAni grown within these ormosil-modified electrodes. Cyclic voltammograms of PAni synthesized within ormosil matrices at various scan rates are shown in Fig. 4, Four redox couples were observed in case of polyaniline formed through prussian blue. The 2nd and 3rd redox couples merged together in single broad peak. The low potential redox peak at 0.2 V can be attributed to a quasi-reversible reaction of leucoemeraldine oxidation to protoemeraldine, and the 2nd and 3rd redox peaks are attributable to oxidation of the later to emeraldine and subsequent oxidation to nigraniline respectively. The later peak suggests the formation of benzoquinone/hydroquinone couple in the acid electrolyte. The highest oxidation potential peak at 0.8 V is attributed to oxidation of nigraniline to pernigraniline and corresponding reduction of the later at 0.7 V[18]. All redox peaks were well resolved and defined in both systems, and the film formation was found to be better than in case of bare ATO and ormosil without prussian blue.

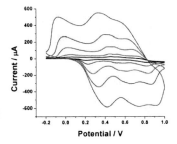

Fig. 4. Cyclic voltammetry of PAni synthesized in ATO modified with ormosil containing prussian blue at 0.01V/s, 0.02 V/s, 0.05 V/s, 0.10 V/s and 0.02 V/s in 1 M HCl between -0.2 V to 1.0 V versus Ag/AgCl.

Spectro-Electrochemistry of PAni grown within Ormosil Matrix

Another remarkable finding on the variation in electrochromic properties of PAni in presence of prussian blue was evaluated from spectro-electrochemical investigation of PAni grown within ormosil matrix. The results recorded are shown in Fig. 5. The observation shown in these figures provides valuable information on; (a) spectroscopy of redox mediator and polyaniline, (b) variation in transmittance as a function of PAni redox behavior, (c) growth pattern of PAni within ormosil network. These observations as shown in figure justified the conclusion; (i) prussian blue-encapsulated ormosil seems to be the better redox matrix for PAni growth under present experimental conditions, (2) the presence of prussian blue within ormosil network introduces patterned PAni.

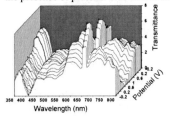

Fig. 5. Spectro-electrochemistry of PAni synthesized in ATO modified with ormosil containing prussian blue at 0.005 V/s in 1 M HCl between -0.2 V to 1.0 V versus Ag/AgCl.

Role of Nano-Structured Domain in Electrocatalysis

The role of nanostructured domain introducing selectivity in catalysis has been attempted in the present investigation with attention to the detection of hydrogen peroxide. Hydrogen peroxide is an important analyte of prime attention as the science of proton and oxygen originates life on the earth. Such analyte does not undergo mediated oxidation and normally introduces over-voltage during direct electrochemistry. Accordingly, the findings on the introduction of electrocatalytic sites from mediated electrochemistry are indeed important requirement for bio-electrochemistry. We report herein the introduction of electrocatalytic sites in nanostructued domain with an example of hydrogen peroxide catalysis involving ormosil encapsulated prussian blue system.The Prussian blue system (PB system) involves a two stage process in ormosil formation. In the first stage, all ormosil's precursors were simultaneously homogenized. The second stage is the mixing of the resultant Prussian blue solution with other ormosil's precursors for generating Prussian blue system. The results on electrochemistry of hydrogen peroxide are shown in Fig. 6a. It should be noted that the particle size of Prussian blue depends on the curing protocol and subsequent mixing of the same with other ormosil precursors for triggering ormosil formation.

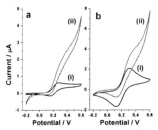

Fig. 6. Cyclic voltammogrames of (a) prussian blue system and (b) Prussian blue-Pd-system before (i) and after (ii) addition of 15 mM H_2O_2 in 0.1 M phosphate buffer, pH 7.0 at 25 °C.

Influence of Modulation via Chemical Reactivity

Another approach for changing the nano-structured domains in ormosil network is to introduce another reaction system prior to initiating ormosil network formation. Such introduction is availed via chemical reactivity of one of the organosilanes participating in ormosil formation. Our previous reports[34, 35] suggested that 3-glycidoxypropyltrimethoxysilane is highly sensitive to the presence of palladium chloride. The epoxide ring of glymo-group is opened by palladium chloride followed by the reduction of palladium (II) into palladium. The reduced palladium is then coordinated with two moieties of glymo-residue[34]. When this reaction product is used for ormosil formation, Pd system is generated. The introduction of palladium within ormosil network not only altered the nano-geometry but also affected the mechanistic approach on the electrochemistry of hydrogen peroxide as well. The electrochemistry of Pd-system before (i) and after (ii) the addition hydrogen peroxide is shown in Fig. 6b. As a consequence, the magnitude of anodic current is decreased as compared to that of PB-system.

Analytical Significance of PB and Pd systems in Electroanalysis of Hydrogen Peroxide

A comparison of apparent electrocatalytic efficiency could be justified from the data on variable sensitivity of hydrogen peroxide analysis made at similar operating potential. The lower the operating potential the higher would be the apparent efficiency of electrocatalysis of the system. We have conducted electroanalysis at +250 mV versus Ag/AgCl and recorded in Fig. 7 that also justifies the selectivity for the hydrogen peroxide oxidation as any conventional electrodes do not show electroanalysis at such a small potential. The calibration curves for hydrogen peroxide detection at the prussian blue-modified electrodes and also at prussian blue along with Pd modified-electrodes were constructed using average currents recorded at three individual films for each concentration point. The results in Fig. 7 show that the Pd system exhibits higher apparent electrocatalytic efficiency on which sensitivity of analysis is calculated to be 0.4 + 0.016 μA/mM. The sensitivity of PB system was found out to be 0.23 + 0.017 μA/mM.

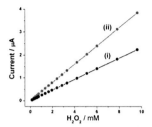

Fig. 7. Calibration curves for hydrogen peroxide analysis at constant potential of 0.25 V versus Ag/AgCl using (i) prussian blue and (i) prussian blue along with Pd in 0.1 M phosphate buffer, pH 7.0 at 25 °C;

CONCLUSION

The application of organically modified silicate (ormosil) matrix in electropolymerization of aniline and in electroanalysis of hydrogen peroxide is reported in the present work. An inorganic redox mediator prussian blue was encapsulated within ormosil film to investigate its role in the polymerization of PAni films formed over the modified electrodes. The results obtained in this study indicate that the electrochemical performance of PAni is better in the presence of redox mediator encapsulated within ormosil film. The redox mediator acts as a continuous conducting relay within the porous ormosil matrix, thereby providing an electronic conduction pathway which improves the process of charge transfer through the matrix. Further, prussian blue and prussian blue along with palladium encapsulated ormosils have been utilized for the electroanalysis of hydrogen peroxide. The significance of these systems is analytically justified on electroanalytical quantification of hydrogen peroxide on these surfaces. The electrochemistry of these systems suggested that desired electrocatalytic sites could be introduced within such matrix based on controlling the nano-geometry of the materials.

REFERENCES
[1]M.S. Cho, Y.H. Cho, H.J. Choi and M.S. Jhon, Synthesis and Electrorheological Characteristics of Polyaniline-Coated Poly(methyl methacrylate) Microsphere: Size Effect, *Langmuir*, **19**, 5875-81 (2003).
[2]K.Aoki, J Chen, Q.Ke, S.P.Armes, and D.P. Randall, Redox Reactions of Polyaniline-Coated Latex Suspensions, *Langmuir*, **19**, 5511-16 (2003).
[3]J.Huang, S.Virji, B.H Weiller, and R.B. Kaner, Polyaniline Nanofibers: Facile Synthesis and Chemical Sensors, *J. Am. Chem. Soc.*, **125**, 314-15 (2003).
[4]V.G. Kulkarni, Tuned conductive coatings from polyaniline, *Synth. Met.*, **71**, 2129-31 (1995).
[5]B. Wessling, and J. Posdorfer, Nanostructures of the dispersed organic metal polyaniline responsible for macroscopic effects in corrosion protection, *Synth. Met.*, **102**, 1400-01 (1999).
[6]P.C. Pandey and G. Singh, Tetraphenylborate doped polyaniline based novel pH sensor and solid-state urea biosensor, *Talanta*, **55**, 773-82 (2001).
[7]P.C. Pandey and G. Singh, Electrochemical Polymerization of Aniline in Proton-Free Nonaqueous Media, *J. Electrochem. Soc.*, **149**, D51-56 (2002).
[8]D.D. Borole, U.R. Kapadi, P.P. Kumbhar and D.G. Hundiwale, Influence of inorganic and organic supporting electrolytes on the electrochemical synthesis of polyaniline, poly(o-toluidine) and their copolymer thin films, *Mater. Lett.*, **56**, 685-91 (2002).

[9]J.R. Santos Jr, J.A. Malmonge, A.J.G. Conceicgo Silva, A.J. Motheo, Y.P. Mascarenhas, and L.H.C. Mattoso, Characteristics of polyaniline electropolymerized in camphor sulfonic acid, *Synth. Met.*, **69**, 141-42 (1995).

[10]Z.J. Ling, Z.X. Gang, X. Fang, and H.F. Ping, Effect of polar solvent acetonitrile on the electrochemical behavior of polyaniline in ionic liquid electrolytes, *J. Colloid Interface Sci.*, **287** 67-(2005).

[11]K. Luo, N. Shi, and C. Sun, Thermal transition of electrochemically synthesized polyaniline, *Polym. Degrad. Stab.*, **91**, 2660-64 (2006).

[12]A.A. Nekrasov, V.F. Ivanov, O.L. Gribkova, and A.V. Vannikov, Voltabsorptometric study of "structural memory" effects in polyaniline, *Electrochim. Acta*, **50**, 1605-13 (2005).

[13]R. Prakash, Electrochemistry of polyaniline: Study of the pH effect and electrochromism, *J. Appl. Polym. Sci.*, **83**, 378-85 (2002).

[14]A.T. Özyılmaz, M. Erbil, and B. Yazıcı, The electrochemical synthesis of polyaniline on stainless steel and its corrosion performance, *Current Appl. Phys.*, **6**, 1-9 (2006).

[15]J. Widera and J.A. Cox, Electrochemical oxidation of aniline in a silica sol–gel matrix, *Electrochem. Commun.*, **4**, 118-22 (2002).

[16]J. Widera, A.M. Kijak, D.V. Ca, G.E. Pacey, R.T. Taylor, H. Perfect, and J.A. Cox, The influence of the matrix structure on the oxidation of aniline in a silica sol–gel composite, *Electrochim. Acta*, **50**, 1703-09 (2005).

[17]B.R Mattes, E.T. Knobbe, P.D. Fuqua, F. Nishida, E.-W. Chang, B.M. Pierce, B. Dunn, and R.B. Kaner, Polyaniline sol-gels and their third-order nonlinear optical effects, *Synth. Met.*, **43**, 3183-87 (1991).

[18]Y. Wei, J.-M. Yen, D. Jin, X. Jia, J. Wang, G.-W. Jang, C. Chen, and R.W. Gumbs, Composites of Electronically Conductive Polyaniline with Polyacrylate-Silica Hybrid Sol-Gel Materials, *Chem. Mater.*, **7**, 969-74 (1995).

[19]R.J.P. Corriu, J.J.E. Moreau, P. Thepot, M.W.C. Man, C. Chorro, J.-P. L_ere-Porte, and J.-L. Sauvajol, Trialkoxysilyl Mono-, Bi-, and Terthiophenes as Molecular Precursors of Hybrid Organic-Inorganic Materials, *Chem. Mater.*, **6**, 640-49 (1994).

[20]C. Sanchez, B. Alonso, F. Chapusot, F. Ribot, and P.J. Audebert, Molecular design of hybrid organic-inorganic materials with electronic properties *J. Sol-Gel Sci. Technol.*, **2**, 1-3 (1994).

[21]G.-W. Jang, C. Chen, R.W. Gumbs, Y. Wei, and J.-M. Yeh, Large-Area Electrochromic Coatings, *J. Electrochem. Soc.*, **143**, 2591-96 (1996).

[22]M.M. Varghese, K. Ramanathan, S.M. Ashraf, M.N. Kamalasanan, B.D. Malhotra, Electrochemical Growth of Polyaniline in Porous Sol–Gel Films, *Chem. Mater.*, **8**, 822-24 (1996).

[23]S. das Neves, S.I. Cordoba de Torresi, R.Ap. Zoppi, Template synthesis of polyaniline: a route to achieve nanocomposites, *Synth. Met.*, **101**, 754-55 (1999).

[24]J.-D. Qiu, H.-Z. Peng, R.-P. Liang, J. Li, and X.-H. Xia, Synthesis, Characterization, and Immobilization of Prussian Blue-Modified Au Nanoparticles: Application to Electrocatalytic Reduction of H_2O_2, *Langmuir*, **23**, 2133–37 (2007).

[25]A.A. Karyakin, O.V. Gitelmacher, and E.E. Karyakina, Prussian Blue-Based First-Generation Biosensor. A Sensitive Amperometric Electrode for Glucose, *Anal. Chem.*, **67**, 2419–23 (1995).

[26]F. Ricci, and G. Palleschi, Sensor and biosensor preparation, optimisation and applications of Prussian Blue modified electrodes, *Biosens. Bioelectron.*, **21**, 389–407 (2005).

[27]P.C. Pandey, B. Singh, R.C. Boro, and C.R. Suri, Chemically sensitized ormosil-modified electrodes—Studies on the enhancement of selectivity in electrochemical oxidation of hydrogen peroxide, *Sens. Actuators B*, **122**, 30–41 (2007).

[28]P.C. Pandey and B.C. Upadhyay, Studies on differential sensing of dopamine at the surface of chemically sensitized ormosil-modified electrodes, *Talanta*, **67**, 997–1006 (2005).

[29]P.C. Pandey, B.C. Upadhyay, and A.K. Upadhyay, Differential selectivity in electrochemical oxidation of ascorbic acid, *Anal. Chim. Acta*, **523**, 219–223 (2004).

[30]P.C. Pandey, B.C. Upadhyay, and A.K. Upadhyay, Electrochemical sensors based on functionalized ormosil-modified electrodes – role of ruthenium and palladium on the electrocatalysis of NADH and ascorbic acid, *Sens. Actuators B*, **102**, 126–131 (2004).

[31]P.C. Pandey, S. Upadhyay, and H.C. Pathak, A New Glucose Biosensor Based on Sandwich Configuration of Organically Modified Sol-Gel Glass, *Electroanalysis*, **11**, 59–64 (1999).

[32]P.C. Pandey, S. Upadhyay, H.C. Pathak, and C.M.D. Pandey, Studies on Ferrocene Immobilized Sol-Gel Glasses and Its Application in the Construction of a Novel Solid-State Ion Sensor, *Electroanalysis* **11**, 950-56 (1999).

[33]P.C. Pandey, S. Upadhyay, H.C. Pathak, I. Tiwari, and V.S. Tripathi, Studies on Glucose Biosensors Based on Nonmediated and Mediated Electrochemical Oxidation of Reduced Glucose Oxidase Encapsulated Within Organically Modified Sol-Gel Glasses, *Electroanalysis*, **11**, 1251–1258 (1999)

[34]P.C. Pandey, S. Upadhyay, and S. Sharma, Functionalized Ormosils-Based Biosensor, *J. Electrochem. Soc.*, **150**, H85-92 (2003).

[35]P.C. Pandey, S. Upadhyay, I. Tiwari, and S. Sharma, A Novel Ferrocene-Encapsulated Palladium-Linked Ormosil-Based Electrocatalytic Biosensor. The Role of the Reactive Functional Group, *Electroanalysis*, **13**, 1519–1527 (2001).

[36]P.C. Pandey, S. Upadhyay, N.K. Shukla, and S. Sharma, *Biosens. Bioelectron.*, **18**, 1257–1268 (2003).

[37]P.C. Pandey, S. Upadhyay, I. Tiwari, and V.S. Tripathi, An ormosil-based peroxide biosensor – a comparative study on direct electron transport from horseradish peroxidase, *Sens. Actuators B*, **72**, 224–232 (2001).

INDIVIDUAL METAL OXIDE NANOWIRES IN CHEMICAL SENSING: BREAKTHROUGHS, CHALLENGES AND PROSPECTS

J. D. Prades[a,b,*], R. Jimenez-Diaz[b], F. Hernandez-Ramirez[a], A. Cirera[b], A. Romano-Rodriguez[b], S. Mathur[c], J. R. Morante[a,b]

[a] Institut de Recerca en Energia de Catalunya (IREC), Barcelona, Spain
[b] IN²UB/XaRMAE, Departament d'Electrònica, Universitat de Barcelona, Barcelona, Spain
[c] Department of Inorganic Chemistry, University of Cologne, Germany

ABSTRACT

Single-crystalline semiconductor metal oxide nanowires exhibit novel structural and electrical properties attributed to their reduced dimensions, well-defined geometry and the negligible presence of grain boundaries and dislocations in their inside. This favors direct chemical transduction mechanisms at their surfaces upon exposure to gas molecules, making them promising active device elements for a new generation of chemical sensors. Furthermore, metal oxide nanowires can be heated up to the optimal operating temperature for gas sensing applications with extremely low power consumption due to their small mass, giving rise to devices more efficient than their nanoparticle-based counterparts. Alternatively, UV light can be used to activate the chemical processes that lead to an electrical response toward gases, even at room temperature. Here, the current status of development of sensors based on individual metal oxide nanowires is surveyed, and the main technological challenges which act as bottleneck to their potential use in real applications are identified.

INTRODUCTION

Metal oxides (MOX) display interesting features that make them suitable for chemical sensing applications[1,2,3]. On the one hand, they represent a low cost alternative to current sensing technologies since MOX can be synthesized in a cost effective way following scalable chemical processes[4]. On the other hand, this is a solid state technology which implies a number of advantages in terms of integration, robustness and packaging, if compared to other technologies (i.e.: electrochemical sensors[5]). In spite of these attractive features, MOX-based gas sensors only represented the 15% of the world market of chemical sensors in 2008[6]; the main reason being that there is still big room for improvement in their performance.

Today, the challenge is going beyond the performances we can obtain with thin or thick films or with layers of nanostructured materials. This specifically means enlarging the response and sensitivity of the sensors, enhancing the stability of the signal, accelerating their response time, lowering the working temperature, reducing their power consumption and improving the selectivity of MOX[7]. In the last years, nanowires have revealed themselves as a promising alternative to implement innovative MOX based gas sensors with better performance[8]. Here, we present a short overview on how nanowires, and specifically individual MOX nanowires, can help to improve all the previous issues.

RESULTS AND DISCUSSION

Sensor Response

Gas sensing with MOX essentially is a surface phenomenon. Chemicals interact with the surface of MOXs trapping or releasing charges, depending on their oxidizing or reducing nature. This surface charge distribution creates an electric field near the surface that bend the electronic bands in the outer shell of the material producing resistance modulation, being this the electrical parameter acquired in conventional conductometric gas sensors[9]. Therefore, it is expected that materials with higher surface-to-volume ratio, like nanomaterials will display larger responses towards gases. To corroborate the previous assertion, Figure 1 shows the response toward the same amount of gas of devices fabricated with different nanomaterials with shrinking radius (r) [10-12]. As of today, ultra-small spherical nanoparticles (r < 15 nm) provide larger responses than nanowires because technical limitations in current nanofabrication methods hamper the integration of ultra-thin nanowires (r < 15nm) in sensors based on individual nanowires[13]. Nevertheless, trends in Figure 1 also indicate that similar responses could be achieved with both kinds of nanoparticles in equivalent surface-to-volume conditions.

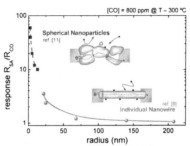

Figure 1. Conductometric response of SnO_2 nanosensors at 300°C toward 800 ppm of CO diluted in synthetic air (SA). The response values reached with ultra-small spherical nanoparticles[12] are compared with the ones obtained with individual nanowires[10], at their present stage of development. R_{SA} stands for the resistance value in SA. R_{CO} stands for the resistance in CO.

Signal Stability

Besides the magnitude of the response, the stability of the signal is also important to guarantee the reliability and the repeatability of the gas measurements. Compared with other nanomaterials (i.e.: sphere-like nanoparticles), nanowires essentially are single crystalline materials enclosed in well defined and stable surfaces[14, 15]. The crystalline quality is important in terms of the electrical transport but, from the gas sensing point of view, the superior stability of their surfaces turns into a better stability, reversibility and reproducibility in their response[10] (see an example in Figure 2). It is noteworthy that this feature is specific of nanowires and it is not available in other nanomaterials.

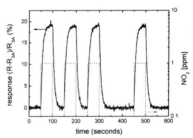

Figure 2: Response of an individual SnO_2 nanowire at 175°C to several pulses of NO_2 diluted in synthetic air (SA). The response of the nanowire-based sensor is stable in time, repeatable and fully reversible. R_{SA} is the initial value of the resistance in SA. R stands for the resistance values during the test.

Response Time

Compared with other approaches, such as porous films of MOX particles, nanowires have intrinsic advantages as far as the response time is concerned. But, nanowires must be used individually to exploit these advantages. It can be easily demonstrated that the relative change in the resistance of a bundle of (identical) nanowires is the same than the relative change in the resistance of each one of the nanowires in the bundle[8]. Therefore every single nanowire encodes all the sensing information and it is able to provide all the sensing response by itself. In addition to this, if we use only one nanowire we will be avoiding other interfering effects, such as the diffusion of gases through the pores and cavities of the bundle that slows down the response time of the sensors[8]. Figure 3 shows an example of the response towards the same amount of gas of an individual nanowire compared to the response obtained with a device based on a bundle of the same wires: the magnitude of the response is comparable but the dynamic response of the individual nanowire is much faster.

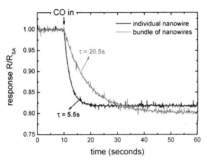

Figure 3: Response transient of a sensor based on one single nanowire and a sensor based on a bundle of them to [CO] = 50 ppm at 300°C. The response of the former device is much faster. R_{SA} is the initial value of the resistance in SA. R stands for the resistance values during the test.

Besides, it has been recently demonstrated that the low thermal inertia of one single nanowire can be exploited to operate these devices in ultra-fast pulsed mode with the help of the self-heating principle (see more details in section "Power Consumption")[16]. Without any systematic optimization, this novel approach already provides thermal responses in the range of 10 ms and delivers response and recovery times to gases of only few tens of milliseconds. This is faster than the times reported for porous film based sensors[17] and it is very close to the ultimate limits in the response times, which are imposed by the chemical kinetics. [18]

Working Temperature

In order to obtain significant responses to gases, it is necessary to heat MOXs up to few hundreds of degrees Celsius[2]. This thermal energy is required to activate the chemical and electrical interactions between gaseous molecules and the atoms at the surface of MOXs[19, 20]. The need of heating limits the utilization of MOX-based gas sensors in certain conditions, such as explosive or flammable environments.

It is well known that the energy of the photons can also be used to activate similar reactions. In recent years there have been several reports[21-26] on the photo-activation of MOX gas sensors. These works demonstrated that ultraviolet illumination can be used to enhance the response of MOXs toward a variety of gases, like CO, NO_2, O_3, H_2 and alkanes, even at room temperature. However, the precise role played by photons in the photo-activation of the chemical transduction processes at the surface of MOXs is still unclear, because the origin of the response is complex to analyze in devices based on polycrystalline or porous films[27]. This fact hampered the systematic improvement and optimization of this technology and impeded lowering the working temperature of MOX.

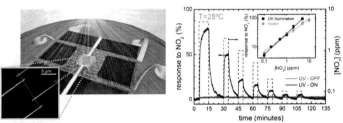

Figure 4: Left: image of a sensor based on one single nanowire under illumination. Right: Response of the sensor toward several pulses of NO$_2$ at room temperature (T = 25°C): UV illumination clearly enhances the response. Inset: comparison of the response of the same sensor operated with conventional heating and with UV illumination. Equivalent responses can be obtained.

Nowadays, the utilization of individual metal oxide nanowires for gas sensing applications offers an advantageous experimental scenario to gain deeper insight into these issues[8]. Firstly, the high surface-to-volume ratio of these nanostructures and their excellent crystalline quality provide enhanced and highly reproducible transduction properties[28], when compared to their microcrystalline counterparts. Second, the interpretation of the resistance modulation mechanism in individual nanowires is straightforward[9] (in contrast with the complex and random distribution of intergrain barriers that modulate the response in polycrystalline materials[27]). Third, individual nanowires, with typical diameter below 100 nm, guarantee uniform illumination conditions that cannot be assured in porous thick films. The possibility to use individual nanowires for UV light-activated gas sensing applications was already demonstrated by the exploratory work of Law et al.[29] Recently, we reported that the response of monocrystalline SnO$_2$ nanowires to NO$_2$ at room temperature is a function of the flux and the energy of photons[30]. We also demonstrated that nearly identical responses to those obtained with thermally activated sensors can be achieved by choosing the optimal illumination conditions at room temperature (see Figure 4). These results are also serving to gain a deeper comprehension of this alternative working mechanism[31].

Power Consumption
    The power needed to heat MOXs to the optimum working temperatures is the main power demand of this technology. In the last years, the use of microtechnologies made possible downscaling the sensors and therefore reducing their power consumption[32]. State-of-the-art devices apply microheaters to thermalize small amounts of sensing material with power requirements of few hundreds of mW[33]. In spite of the improvements achieved so far, current power values still hinder the development of very attractive applications, such as fully autonomous sensor networks for distributed gas detection.
    Recently, we have demonstrated that the current applied to individual nanowires in conductometric operation can be used to dissipate enough electrical power to self-heat the tiny mass of the nanowire and thus, reach the optimum working conditions for the detection of chemicals, avoiding the need of external heaters[34, 35]. An important consequence of the integration of the heater in the

sensing nanomaterial is the dramatic reduction of the heating volume and consequently, of the power requirements. According to our preliminary measurements, these nanowires could operate with less than 20 μW to both bias and heat them[34]. This estimation, which is more than two orders of magnitude lower than the power consumption of the microheaters, compelled us to combine the self-heating principle with energy harvesting technologies to develop self-powered chemical sensor systems without the need of battery replacement[36,37]. These proof-of-concept devices illustrate how individual nanowires enable new possibilities which were just unfeasible with previous technologies.

It is noteworthy that the self-heating approach can be used in pulsed operation mode (see section "Response Time") with different duty cycles to tailor at will and further reduce the power requirements[38].

Selectivity

The selectivity of MOXs toward different gases is poor and its improvement is still a challenge[39]. Recent findings have shown that this aim can only be achieved via a deeper comprehension of the interaction mechanisms between the gas molecules and the active surfaces[1, 40], either clean or decorated with functionalizing agents[41]. As previously stated, individual nanowires are offering an outstanding and well defined experimental scenario to advance in this understanding[8].

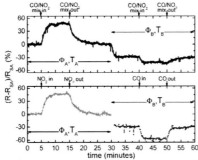

Figure 5: Fundamental findings for a gas separation algorithm. Under the optimum light flux and temperature conditions (A conditions), the response of the sensor is immune to the presence of CO. These CO immune conditions serve to obtain accurate and selective estimations of the NO_2 concentration. At different illumination and temperature (B conditions), the previous estimations can be used to correct the reading of the sensor exposed to a mixture of CO and NO_2 and deduce the concetration of CO. This two-stage algorithm was successfully implemented with sensors based on individual nanowires[41]. R_{SA} is the initial value of the resistance in SA. R stands for the resistance values during the test.

Besides, recent findings demonstrate that heat and illumination can be combined in nanowires to produce opposite responses toward different gases[42]. Figure 5 shows an example of this strategy. At 100°C and under UV illumination, SnO_2 nanowires display a positive response to oxidizing gases (NO_2). In these conditions heat and light activate two opposite mechanisms in the response toward reducing gases (CO) in a way that the sensor is almost insensitive to these compounds. Thus, it is possible to obtain accurate estimates of the concentration of oxidizing gases without the interfering effect of the reducing ones. This opens the door to the development of advanced gas separation algorithms for a superior selectivity. It is noteworthy that this unusual behavior can only be achieved by fine tuning the external excitation sources. The small size and simple layout of one devices based on one single nanowire facilitates this task.

Challenges in Nanofabrication
In spite of the attractive properties described so far, there exist important technical difficulties that hinder the utilization of individual nanowires in real live sensor devices. It has been demonstrated that it is possible to manipulate these materials and to fabricate devices with them, but only at laboratory scale. To date, there are not any clear methodologies to produce devices based on individual nanowires in scalable and cost effective ways. Nevertheless, there are increasing efforts being undergone to palliate this situation such as the development of self-assembly strategies like dielectrophoresis[43]. More details on the challenges in nanofabrication can be found elsewhere[13].

CONCLUSION
The high surface-to-volume ratio of nanomaterials makes them suitable to enhance the response in gas sensing applications. Amongst the variety of MOX nanoparticles that can be synthesized nowadays, nanowires display some advantageous features. Their superior surface stability turns into a better stability and reversibility in the response. If used individually, nanowires provide an outstanding experimental situation to understand better the chemical processes at the surface of MOX that led to the electrical response to gases. In addition, this simple configuration avoids many interfering effects, such as the gas diffusion processes which are typical of porous layers and worsen the dynamic response. Moreover their reduced dimensions make possible activating the chemical processes with self-heating or with UV illumination in a precise and controlled way. These strategies enable new and previously unexpected features such as ultra low power consumption and room temperature operation capability. All these facts show how individual nanowires offer an attractive opportunity to go beyond current performances in MOX sensors technology.

FOOTNOTES
* Corresponding author. E-mail: dprades@el.ub.es

REFERENCES
[1]   G. Korotcenkov, Metal Oxides for Solid-State Gas Sensors: What Determines Our Choice?, *Materials Science and Engineering B-Solid State Materials for Advanced Technology*, **139**, 1-23 (2007).
[2]   N. Yamazoe, and K. Shimanoe, New Perspectives of Gas Sensor Technology, *Sensors and Actuators B Chemical*, **138**, 100-7 (2009).
[3]   E. Comini, Metal Oxide Nano-Crystals for Gas Sensing, *Analytica Chimica Acta*, **568**, 28-40 (2006).
[4]   M. Batzill, and U. Diebold, The Surface and Materials Science of Tin Oxide, *Progress in Surface Science*, **79**, 47-154 (2005).
[5]   G. Korotcenkov, S. Do Han, and J. R. Stetter, Review of Electrochemical Hydrogen Sensors, *Chemical Reviews*, **109**, 1402-33 (2009).
[6]   Global Industry Analysts "Sensors: A Global Strategic Business Report" (2008).
[7]   G. Korotcenkov, Practical Aspects in Design of One-Electrode Semiconductor Gas Sensors: Status Report, *Sensors and Actuators B-Chemical*, **121**, 664-78 (2007).
[8]   F. Hernandez-Ramirez, J. D. Prades, R. Jimenez-Diaz, T. Fischer, A. Romano-Rodriguez, S. Mathur, and J. R. Morante, On the Role of Individual Metal Oxide Nanowires in the Scaling Down of Chemical Sensors, *Physical Chemistry Chemical Physics*, **11**, 7105-10 (2009).
[9]   F. Hernandez-Ramirez, J. D. Prades, A. Tarancon, S. Barth, O. Casals, R. Jimenez-Diaz, E. Pellicer, J. Rodriguez, J. R. Morante, M. A. Juli, S. Mathur, and A. Romano-Rodriguez, Insight into

the Role of Oxygen Diffusion in the Sensing Mechanisms of SnO_2 Nanowires, *Advanced Functional Materials*, **18**, 2990-4 (2008).

[10] F. Hernandez-Ramirez, A. Tarancon, O. Casals, J. Arbiol, A. Romano-Rodriguez, and J. R. Morante, High Response and Stability in CO and Humidity Measures Using a Single SnO_2 Nanowire, *Sensors and Actuators B-Chemical*, **121**, 3-17 (2007).

[11] N. Yamazoe, and K. Shimanoe, Roles of Shape and Size of Component Crystals in Semiconductor Gas Sensors, *Journal of the Electrochemical Society*, **155**, J85-J92 (2008).

[12] N. Yamazoe, and K. Shimanoe, Roles of Shape and Size of Component Crystals in Semiconductor Gas Sensors, *Journal of the Electrochemical Society*, **155**, J93-J8 (2008).

[13] F. Hernandez-Ramirez, J. D. Prades, S. Barth, A. Romano-Rodriguez, S. Mathur, A. Tarancón, O. Casals, R. Jimenez-Diaz, J. Rodriguez, E. Pellicer, M. A.l Juli, T. Andreu, S. Estrade, E. Rossinyol, J. R. Morante, Fabrication of Nanodevices Based on Individual SnO_2 Nanowires and Their Electrical Characterization, in Metal Oxide Nanostructures and Their Applications (Ed. A. Umar and Y.-B. Hahn), **3** , 1-28 (2010).

[14] S. Mathur, S. Barth, H. Shen, J. C. Pyun, and U. Werner, Size-Dependent Photoconductance in SnO_2 Nanowires, *Small*, **1**, 713-7 (2005).

[15] S. Mathur, and S. Barth, Molecule-Based Chemical Vapor Growth of Aligned SnO_2 Nanowires and Branched SnO_2/V_2O_5 Heterostructures, *Small*, **3**, 2070-5 (2007).

[16] J. D. Prades, R. Jimenez-Diaz, F. Hernandez-Ramirez, J. Pan, A. Romano-Rodriguez, S. Marthur, and J. R. Morante, Direct Observation of the Gas-Surface Interaction Kinetics in Nanowires through Pulsed Self-Heating Assisted Conductometric Measurements, *Applied Physics Letters*, **95**, 053101 (2009).

[17] T. Kida, T. Kuroiwa, M. Yuasa, K. Shimanoe, and N. Yamazoe, Study on the Response and Recovery Properties of Semiconductor Gas Sensors Using a High-Speed Gas-Switching System, *Sensors and Actuators B-Chemical*, **134**, 928-33 (2008).

[18] A. Helwig, G. Müller, G. Sberveglieri, and G. Faglia, Gas Response Times of Nano-Scale SnO_2 Gas Sensors as Determined by the Moving Gas Outlet Technique, *Sensors and Actuators B: Chemical*, **126**, 174-80 (2007).

[19] J. D. Prades, A. Cirera, J. R. Morante, J. M. Pruneda, and P. Ordejon, Ab Initio Study of NO_x Compounds Adsorption on SnO_2 Surface, *Sensors and Actuators B-Chemical*, **126**, 62-7 (2007).

[20] J. D. Prades, A. Cirera, and J. R. Morante, First-Principles Study of NO_x and SO_2 Adsorption onto SnO_2(110), *Journal of the Electrochemical Society*, **154**, H675-H80 (2007).

[21] J. Saura, Gas-Sensing Properties of SnO_2 Pyrolitic Films Subjected to Ultrviolet Radiation, *Sensors and Actuators B: Chemical*, **17**, 211-4 (1994).

[22] E. Comini, A. Cristalli, G. Faglia, and G. Sberveglieri, Light Enhanced Gas Sensing Properties of Indium Oxide and Tin Dioxide Sensors, *Sensors and Actuators B: Chemical*, **65**, 260-3 (2000).

[23] E. Comini, G. Faglia, and G. Sberveglieri, Uv Light Activation of Tin Oxide Thin Films for NO_2 Sensing at Low Temperatures, *Sensors and Actuators B: Chemical*, **78**, 73-7 (2001).

[24] S. Mishra, C. Ghanshyam, N. Ram, R. P. Bajpai, and R. K. Bedi, Detection Mechanism of Metal Oxide Gas Sensor under UV Radiation, *Sensors and Actuators B: Chemical*, **97**, 387-90 (2004).

[25] C.-H. Han, D.-W. Hong, S.-D. Han, J. Gwak, and K. C. Singh, Catalytic Combustion Type Hydrogen Gas Sensor Using TiO_2 and UV-Led, *Sensors and Actuators B: Chemical*, **125**, 224-8 (2007).

[26] B. P. J. de Lacy Costello, R. J. Ewen, N. M. Ratcliffe, and M. Richards, Highly Sensitive Room Temperature Sensors Based on the UV-Led Activation of Zinc Oxide Nanoparticles, *Sensors and Actuators B: Chemical*, **134**, 945-52 (2008).

[27] N. Barsan, D. Koziej, and U. Weimar, Metal Oxide-Based Gas Sensor Research: How To?, *Sensors and Actuators B-Chemical*, **121**, 18-35 (2007).

[28]   M. Law, J. Goldberger, and P. Yang, Semiconductor Nanowires and Nanotubes, *Annual Review of Materials Research*, **34**, 83-122 (2004).

[29]   M. Law, H. Kind, B. Messer, F. Kim, and P. Yang, Photochemical Sensing of NO_2 with SnO_2 Nanoribbon Nanosensors at Room Temperature, *Angewandte Chemie International Edition*, **41**, 2405-8 (2002).

[30]   J. D. Prades, R. Jimenez-Diaz, F. Hernandez-Ramirez, S. Barth, A. Cirera, A. Romano-Rodriguez, S. Mathur, and J. R. Morante, Equivalence between Thermal and Room Temperature UV Light-Modulated Responses of Gas Sensors Based on Individual SnO_2 Nanowires, *Sensors and Actuators B: Chemical*, **140**, 337-41 (2009).

[31]   J. D. Prades, M. Manzanares, R. Jimenez-Diaz, F. Hernandez-Ramirez, J. Pan, A. Cirera, A. Romano-Rodriguez, S. Mathur, and J. R. Morante, On the Role of the Illumination Conditions in the Response Towards Oxidizing Gases of Room Temperature Sensors Based on Individual SnO_2 Nanowires, in preparation (2009).

[32]   J. Puigcorbé, A. Vilà, and J. R. Morante, Thermal Fatigue Modeling of Micromachined Gas Sensors, *Sensors and Actuators B: Chemical*, **95**, 275-81 (2003).

[33]   F. Hernandez-Ramirez, J. D. Prades, A. Tarancon, S. Barth, O. Casals, R. Jimenez-Diaz, E. Pellicer, J. Rodriguez, M. A. Juli, A. Romano-Rodriguez, J. R. Morante, S. Mathur, A. Helwig, J. Spannhake, and G. Mueller, Portable Microsensors Based on Individual SnO_2 Nanowires, *Nanotechnology*, **18**, 5 (2007).

[34]   J. D. Prades, R. Jimenez-Diaz, F. Hernandez-Ramirez, S. Barth, A. Cirera, A. Romano-Rodriguez, S. Mathur, and J. R. Morante, Ultralow Power Consumption Gas Sensors Based on Self-Heated Individual Nanowires, *Applied Physics Letters*, **93**, 3 (2008).

[35]   J. D. Prades, R. Jimenez-Diaz, F. Hernandez-Ramirez, S. Barth, J. Pan, A. Cirera, A. Romano-Rodriguez, S. Mathur, and J. R. Morante, An Experimental Method to Estimate the Temperature of Individual Nanowires, *International Journal of Nanotechnology*, **6**, 860 - 9 (2009).

[36]   J.D. Prades, R. Jimenez-Diaz, F. Hernandez-Ramirez, A. Cirera, A. Romano-Rodriguez, J.R. Morante, Individual Nanowire Chemical Sensor System Self-Powered with Energy Scavenging Technologies, Proceedings of the Transducers2009 Conference, Denver (2009).

[37]   F. Hernandez-Ramirez, J. D. Prades, J. R. Morante, A. Cirera, A. Romano-Rodríguez "Sensor para gases energéticamente eficiente, red de sensores y procedimiento de medida empleando dicho sensor", Spanish patent application No. P200900334 (2 February 2009).

[38]   J.D. Prades, R. Jimenez-Diaz, F. Hernandez-Ramirez, A. Cirera, A. Romano-Rodriguez, and J.R. Morante, Harnessing self-heating in nanowires for energy efficient, fully-autonomous and ultra-fast gas sensors, submitted to Sensors and Actuators B: Chemical (2009).

[39]   P. Balaji, and et al., Sensitivity, Selectivity and Stability of Tin Oxide Nanostructures on Large Area Arrays of Microhotplates, *Nanotechnology*, **17**, 415 (2006).

[40]   M. Epifani, J. D. Prades, E. Comini, E. Pellicer, M. Avella, P. Siciliano, G. Faglia, A. Cirera, R. Scotti, F. Morazzoni, and J. R. Morante, The Role of Surface Oxygen Vacancies in the NO_2 Sensing Properties of SnO_2 Nanocrystals, *Journal of Physical Chemistry C*, **112**, 19540-6 (2008).

[41]   C. Di Valentin, U. Diebold, and A. Selloni, Doping and Functionalization of Photoactive Semiconducting Metal Oxides, *Chemical Physics*, **339**, vii-viii (2007).

[42]   J. D. Prades, F. Hernandez-Ramirez, J. R. Morante, A. Cirera, A. Romano-Rodríguez "Sensor de la concentración de un gas, red de sensores y procedimientos de medida de la concentración a partir de dichos sensores", Spanish patent application No. P200930050 (7 April 2009).

[43]   S. Kumar, S. Rajaraman, R. A. Gerhardt, Z. L. Wang, and P. J. Hesketh, Tin Oxide Nanosensor Fabrication Using AC Dielectrophoretic Manipulation of Nanobelts, *Electrochimica Acta*, **51**, 943-51 (2005).

PREPARATION AND THEIR MECHANICAL PROPERTIES OF Al_2O_3/Ti COMPOSITE MATERIALS

Enrique Rocha-Rangel
Departamento de Ing. Metalúrgica, ESIQIE-IPN,
UPALM, Av. IPN S/N, San Pedro Zacatenco, México, D. F. 07738,

Elizabeth Refugio-García, José G. Miranda-Hernández
Departamento de Materiales, Universidad Autónoma Metropolitana
Av. San Pablo # 180, Col Reynosa-Tamaulipas, México, D. F., 02200

Eduardo Térres-Rojas
Laboratorio de Microscopía Electrónica de Ultra Alta Resolución, IMP
Eje Central Lazara Cárdenas # 152, Col San Bartolo Atepehuacan, México, D. F., 07730

Sebastián Díaz de la Torre
CIITEC-IPN
Cerrada de Cecati S/N, Santa Catarina, México D. F. 02250

ABSTRACT
The effect of different titanium additions (0.5, 1, 2 and 3 wt %) and the milling intensity (4 and 8 h) in a Symoloyer mill on the microstructure and fracture toughness of Al_2O_3-based composite materials is analyzed in this work. After high energy milling of Al_2O_3-Ti powder mixtures, it was obtained small particles with main size of about 150 nanometers. The microstructures observed by SEM present the formation of a small and fine metallic net inside the ceramic matrix. From fracture toughness measurements made by the indentation fracture method, it was observed that when the concentration of titanium increases, the fracture toughness of the composites also increases. Enhance in the fracture toughness is due to the bridging formation by the metal in the ceramic matrix. On the other hand, with milling times of 8 h combined with sintering temperatures of 1500 °C during 1 h, they were obtained well consolidated bodies.

INTRODUCTION
In recent times, interpenetrating composites between ceramics and metals (cermets) have been considered to take advantage of the best properties of both phases. High wear resistance is achieved from ceramic/metal microstructures because of the high hardness and high wear resistance of the ceramic fraction in the composite. The metallic fraction increases the fracture toughness of the composite, which improves its damage tolerance. Interpenetrating composites have an advantage over other materials, because the homogeneous distribution of the metal in the ceramic matrix provides dimensional stability at high temperatures[1-5]. Interpenetrating cermets can be fabricated by a number of techniques such as: direct oxidation of a metal[6], metal infiltration of a ceramic perform[7-8], reactive metal penetration[9-10], by hot pressing[11-13] and by thermal spray processing[14]. However, most of these processes are costly, present low productivity and they are complex in their procedures. Therefore, simple and cheaper processes are now in development for the production of high amounts of cermets. High-energy ball milling combined with pressureless sintering can be a substitute low-cost method for the production of cermets, and conventional powder-techniques can be applied for forming and densification. Milling can be performed under vacuum, air or inert atmosphere as well as under dry or wet conditions, using an organic milling agent[15], in addition, it allows for variation of the metallic phase or the ceramic volume fraction in a wide range. In this way, the properties of the composite can be modified to suit the desired purpose.

111

EXPERIMENTAL PROCEDURE

The starting materials were Al$_2$O$_3$ powder (99.9 %, 1 μm, Sigma, USA) and Ti powder (99.9 %, 1-2 μm, Aldrich, USA). The final titanium contents in the produced composites were 0.5, 1, 2 and 3 vol. %. Powder blends of 20 g were prepared in high energy mill (simoloyer) with ZrO$_2$ media, the rotation speed of the mill was of 400 rpm, two different milling times were studied (4 and 8 h). With the milled powder mixture, green cylindrical compacts with size of 2 cm diameter and 0.2 cm thickness were fabricated by uniaxial pressing, using 270 MPa pressure. Then pressureless sintering was performed at two different temperatures (1500 and 1600 °C) during 1 h, under argon flux of 10 cm^3/min. Densities of sintered specimens were determined by using Archimedes principle. The microstructure was observed by scanning electron microscopy (SEM). The hardness of samples was evaluated using Vickers indentation, fracture toughness was estimated by the fracture indentation method[16], (in all cases ten independent measurements per value were carry out).

RESULTS AND DISCUSSION

Density

The results of density measurements for all samples appear in Figure 1. Here it is possible to observe that the effect of the atoms diffusion during the sintering stage causes the densification of all samples. From these values it is considered that some pore formation occur during sintering, the pore formation can be provoked in the sample since the titanium is well known that not wet ceramics[17], due to this there is the presence of some remained porosity in samples. On the other hand, also is observed that the best values of density are reached by samples prepared with powder milled during 8 h and sintered at 1500 °C.

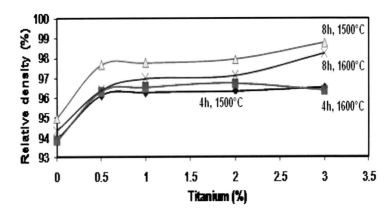

Figure 1. Final density values reached for sintered cermets as a function of titanium content, milling time and sintered temperature.

Hardness
The measured micro-hardness of the samples as a function of titanium content in the composite, milling time and sintered temperature are reported in Figure 2. In this figure it is seen an unusual behavior, because hardness in all samples with 0.5 % Ti present a strong reduction, however for contents of 1 and 2 % Ti the hardness experiment an increment in their values, finally values of hardness for samples with 3 % Ti exhibit intermediate values. This behavior is difficult to explain because it was waited, a decrement in the hardness of samples with the increments of titanium due to the minor hardness of a metallic phase in comparison with the hardness of a ceramic phase.

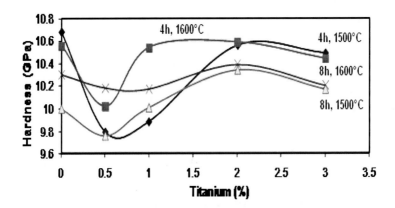

Figure 2. Hardness values of sintered cermets as a function of titanium
content, milling time and sintered temperature.

Fracture toughness
The results of mechanical test show that the fracture toughness is enhanced when in the composites' microstructure there is the presence of a ductile metal. The influence of the milling time and the sintered temperature also have an important effect on the final values of fracture toughness, because samples prepared with powder milled during 8 h and sintered at 1500 °C display the best values of fracture toughness when it is compared with those samples prepared with powder milled during 4 h and sintered at 1600 °C. This can be explained by the finest metallic particle size achieved in samples prepared with powder milled during 8 h and sintered at 1500 °C. Several authors have been reported that enhancements in fracture toughness in cermets may be due to plastic deformation of the metallic phase, which forms crack-bridging ligaments[7, 15, 18].

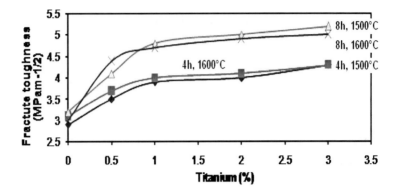

Figure 3. Fracture toughness of sintered cermets as a function of titanium content, milling time and sintered temperature.

Microstructure

Figure 2 shows scanning electron typical micrographs of cermets with 2 and 3 % Ti. Suitably refined and homogeneous microstructures are achieved in all samples. The ligament diameter (titanium) ranges from less than 1 μm and it appear be independent of the amount of titanium in the composite. The grain size of alumina ranges from 3 to 12 μm and it appears to grow up with the increments of Ti in the composite. In these pictures it is possible to observe some porosity in the samples.

Figure 4. Microstructures of sintered cermets as a function of titanium content.
(a) sample with 2% Ti and (b) sample with 3% Ti.

In all samples, the alumina-matrix and reinforcing metals were identified with the help of EDS analysis performed during SEM observations an spectra of this analysis carry out in sample with 3% Ti are show in Figure 5. This spectra show the titanium particles immerse in the ceramic matrix.

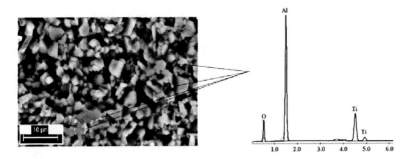

Figure 5. Spectra of EDS analysis performed in sample with 3% Ti, the spectra shows that titanium are the small white particles localized at intergranular position in the matrix.

CONCLUSIONS

- Al$_2$O$_3$/Ti cermets with interpenetrating microstructures can be manufactured by the combination of high energy milling and pressureless sintering of Al$_2$O$_3$ and Ti powders.
- The best process conditions are 8 h high energy milling at 400 rpm and sintering at 1500 °C during 1 h.
- Homogeneous composites with fracture toughness as high as 5.2 MPa·m$^{1/2}$ for a composite with 3 vol % Ti could be obtained and show better damage tolerance than does monolithic Al$_2$O$_3$.
- The refined and homogeneous incorporation of a ductile metal (Ti) in a hard ceramic matrix (Al$_2$O$_3$) improves its fracture toughness.

ACKNOWLEDGMENT

Authors would thank Universidad Autónoma Metropolitana for technical and financial support from 2260235 project and Microscopy laboratory from IMP for technical support.

REFERENCES

[1]C. J. McMahon Jr., *Structural Materials*, Merion Books, 315-348 (2004).
[2]J. K. Wessel, *The Handbook of Advanced Materials*, John Wiley & Sons, 66-88 (2004).
[3]Y. Miyamoto, W. A. Kaysser, B.H Rabin, A. Kawasaki and R.G. Ford (eds.): *Functionally Graded Materials; Design, Processing and Applications*. Kluwer Academic, USA. 1999.
[4]B.H. Rabin and I. Shiota: *Functionally Gradient Materials*, MRS Bull., 14-18 (1995).
[5]O.L. Ighodaro and O.I. Okoli, Fracture Toughness Enhancement for Alumina Systems: A Review. *Int. J. Appl. Ceram. Technol.*, 313-323 (2008).
[6]M. S. Newkirk, A. W. Urquhart, H. R. Zwicker and E. Breval, Formation of Lanxide^{TM} Ceramic Composite Materials, *J. Mater. Res.*, 1, 81-89 (1986).
[7]S. Wu, A. J. Gesing, N. A. Travitzky and N. Claussen, Fabrication and Properties of Al-Infiltrated RBAO-Based Composites, *J. Eur. Ceram. Soc.*, 7, 277-281 (1991).
[8]C. Toy and W. D. Scott, Ceramic-Metal Composite Produced by Melt-Infiltration, *J. Am. Ceram. Soc.*, 73, 97-101 (1990).
[9]S. M. Naga, A. El-Maghraby and A. M. El-Rafei, Properties of Ceramic-Metal Composites Formed by Reactive Metal Penetration, *Am. Cerm. Bull.*, 86, 9301-9307 (2007).

[10]R. E. Loehman, K. Ewsuk and P. Tomsia, Synthesisi of Al₂O₃/Al Composites by Reactive Metal Penetration, *J. Am. Ceram. Soc.*, **79**, 27-32 (1996).

[11]E. Rocha and C. Vilchis, Fabrication and Consolidation of NiAl/Al₂O₃ Composites Using Simultaneously In-Situ Displacement Reactions and Hot Pressing, *in Proceedings of 5th International Conference on High-Temperature Ceramic Matrix Composites – HTCMC5*, American Ceramic Society, 347-352 (2004).

[12]N. Claussen, M. Knechtel, H. Prielipp and J. Rodel, A Strong Variant of Cermets, *Ber. Dtsch. Keram. Ges.* 301-303 (1994).

[13]S.J. Ko, K.H. Min and Y.D. Kim, A study on the fabrication of Al₂O₃/Cu nanocomposite and its mechanical Properties, *Journal of Ceramic Processing Research.* 192-194 (2002).

[14]S. Sampath, H. Herman, N. Shimoda and T. Saito, Thermal Spray Processing of FGM's, *MRS Bull.* 27-31 (1995).

[15]J. Rodel, H. Prielipp, N. Claussen, M. Sternitzke, K. B. Alexander, P. F. Becher and J. H. Schneibel, Ni₃Al/Al₂O₃ Composites with Interpenetrating Networks, *Scr. Met. Mater.*, **33** 843-848 (1995).

[16]A. G. Evans and E. A. Charles, Fracture Toughness Determination by Indentation, *J. Am. Ceram. Soc.*, **59**, 371-372 (1976).

[17]T. Klassen, R. Günther, B. Dickau, F. Gärtner. A. Bartels, R. Bormann and H. Mecking, *J. Am. Ceram. Soc.*, **81**, 2504-2506 (1998).

[18]T. Klassen, R. Gunther, B. Dickau, F. Gartner, A. Bartels, R. Bormann and H. Mecking, Processing and Properties of Intermetallic/ceramics Composites with Interpenetrating Microstructure, *J. Am. Ceram. Soc.*, **81**, 2504-2506 (1998).

BIPHASIC NANO-MATERIALS AND APPLICATIONS IN LIFE SCIENCES: 1D Al/Al2O3
NANOSTRUCTURES FOR IMPROVED NEURON CELL CULTURING

M. Veith*, O. C. Aktas, J. Lee, M. M. Miró, C. K. Akkan

INM – Leibniz Institute for New Materials, Campus D2 2, 66123 Saarbrücken / Germany
K. H. Schäfer, U. Rauch
Fachhochschule Kaiserslautern, Informatik und Mikrosystemtechnik, Campus, 66482 Zweibrücken /
Germany

ABSTRACT
        The development of surfaces which improve the biocompatibility and cell-substrate adhesion is
of great interest in bio-applications especially in neuroscience. In this regard one-dimensional (1D)
Al/Al$_2$O$_3$ nanostructures were synthesized by the chemical vapor deposition (CVD) of [^{t}BuOAlH$_2$]$_2$.
The deposited layers were characterized by scanning electron microscopy (SEM), energy dispersive
spectroscopy (EDX) and X-ray photoemission spectroscopy (XPS). Wetting was analyzed with a
video-contact angle meter. Following the detailed surface characterizations, the neurons isolated from
dorsal root ganglia (DRG) were seeded on Al/Al$_2$O$_3$ coated substrates and the interaction of neurons
with the nanostructured surface was studied. It is observed that neurons form extremely dense
networks on Al/Al$_2$O$_3$ coated substrates compared to planar glass control substrates.

INTRODUCTION
        In recent years nanomaterials are used more and more in biomedical applications especially in
basic and clinical neuroscience. Different nanostructured materials including quantum dots, nanowires
or nanorods are used to study the interaction and stimulation of neurons at the molecular level [1], the
transport of drugs and small molecules [2], neuronal differentiation, neuron regeneration [3] and
communication/signaling mechanism of neurite networks.

        1D structures are employed in neuroscience for a long time. For instance, micropipette
electrodes and micro fabricated electrodes are well established tools to explore the electrical behavior
of individual neuron and neuronic network [4]. Such microstructures are not small enough to explore
neuronal activities at the level of an individual axon. In this context, 1D nanostructures may act as
attractive tools for the coupling of neurons to technical surfaces, for example to improve the adhesion
between neurons and electrode. Lieber et al. showed that a nanowire contacted to a single axon can
measure a conductance peak [5].

        1D nanostructures can be also used for transporting foreign-genes into cells. McKnight et al.
showed that the penetration of DNA-modified carbon nanofibers into living cells provides an efficient
delivery and expression of exogenous genes which reminds of former microinjection methods [6]. On
the other hand such carbon nanofibers are in the form of micro-bunches and an external force is needed
to push the cells on to the tips. For the DNA injection, Hällstrom et al. showed a direct interaction of
cells without an external force by using isolated GaP nanowires [7]. They also presented that neuron
cells survive better on nanowires deposited substrates than planar control surfaces.

        In this present study, we report our first investigations of the interaction of neuron cells with 1D
composite nanostructures. In this preliminary work neurons isolated from rat DRG were seeded on 1D
Al/Al$_2$O$_3$ nanostructures prepared by CVD of [^{t}BuOAlH$_2$]$_2$ to demonstrate the influence of the 1D
topography upon neuronal growth and the formation of neuronal networks.

EXPERIMENTAL

Preparation of 1D Al/Al$_2$O$_3$ Nanostructures

1D Al/Al$_2$O$_3$ nanostructures were synthesized simply by the decomposition of the molecular precursor [^{t}BuOAlH$_2$]$_2$ at elevated temperatures under vacuum. The precursor was synthesized prior to the deposition by a reaction between anhydrous AlCl$_3$ and LiAlH$_4$ and a subsequent reaction with tert-butanol (^{t}BuOH), as described elsewhere [8, 9]. Round shaped glass slides with diameter of 12 mm were employed as the substrates. Glass substrates were cleaned with iso-propanol and dried at 200°C for 30 minutes to remove any organic residue. The substrates were placed on a graphite sample holder in order to be heated by the induction coil using a standard high frequency generator. Figure 1 shows the schematic illustration of the CVD chamber.

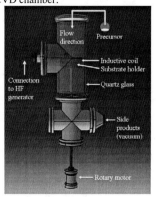

Figure 1: CVD chamber schematic

Before the deposition process, the chamber was evacuated more than 30 minutes until the vacuum reaches the 1~5x10^{-3} mbar level and then flushed with N$_2$ several times. Afterwards the substrates were heated up to 600-650°C and the precursor was flowed into the reaction chamber. The flow rate was controlled manually with a driven valve and a sensitive pressure detector. Following 30 minutes deposition, the precursor valve was closed and substrates were cooled down to room temperature.

Surface Characterization

In order to analyze the morphology of the deposited layers, substrates were firstly coated with gold (Au) using JEOL JFC-1300 fine coater. Then the surfaces of the deposited layers were imaged with SEM (JEOL-JSM-6400F) at an accelerating voltage of 15kV. The surface composition was analyzed with EDX coupled to SEM. In addition to EDX analysis, surface composition was analyzed using a PHI 5600 XPS employing a monochromatic AlKα X-ray prior to cell-culture experiments. Wetting behaviors of the surfaces were tested using a semi-automated contact angle meter (Krues G2).

Cell Culturing and Microscopic Analysis of Cells

Neurons isolated from neonatal Sprague-Dawley rats postnatal day 5 (P5) DGR's were seeded on the top of the deposited layers according to protocol shown previously [10]. After cultivation for 5 days the cells were fixed with formaline (4%) for 10 minutes at room temperature, washed with DBPS (Linaris, Germany) and could be stored in PBS at 8°C in a refrigerator. DRG neurons were stained with a mouse monoclonal antibody against BIII-tubulin (MAB1637, Chemicon, Germany).

Visualization was done with a fluorescent second antibody (AlexaFluor 488 goat anti-mouse IgG, Invitrogen, Germany). The complete analysis was done with an inverted microscope (Olympus IX50, Olympus, Germany). Photos were taken with a ColorView digital camera (Olympus, Germany) and image analysis was done with the analysis software (analysis docu 5.0, Olympus, Germany). Neuron adhesion, morphology and fiber density were inspected visually.

RESULTS

Digital photos given in Figure 2a show coated and non-coated round glass slides. Coated glass slides look black and this may be due to enhanced absorption by the tangled and interpenetrated nanowires. Although the layer looks dark, the transparency is enough to employ a fluorescence microscope in transmission mode. Figure 2b shows the SEM image of randomly grown nanowires at 600 ~ 650°C. Previously we have shown that such nanowires are composed of an inner Al core wrapped with Al_2O_3 shell of a constant molar ratio of Al:Al/Al_2O_3 nanostructures = 1:1 [11]. One can see in a magnified image (Figure 2c) that there are spherical particles at the end of the nanowires which remind of catalyst assisted grown 1D nanostructures [12]. Absence of such a catalyst particle on the substrate indicates a self-catalyst growth. EDX analysis in Figure 2d shows that Al:O ratio is 47.06:52.94 and previously we have shown using TEM that this ratio changes through the cross section of a single nanowire [11] since the core is metallic Al and the shell is Al_2O_3.

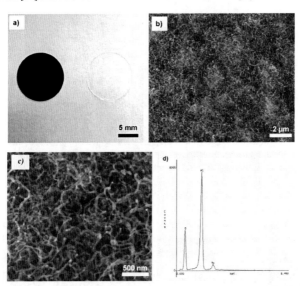

Figure 2. (a) Digital photo of coated (black) and non-coated round shaped glass substrates, SEM images of 1D Al/Al_2O_3 nanostructures at (b) low magnification and (c) high magnification and (d) EDX analysis of 1D Al/Al_2O_3 nanostructures

It is believed that random growth is a consequence of the thermal and gas flow gradients within the chamber. The random grown 1D Al/Al_2O_3 nanostructures yield a highly porous layer. Wetting measurements show that the water droplet exhibits a contact angle below 10° on this porous surface (consistent with a high hydrophilicity). After a few seconds, the water droplet penetrated through the highly porous layer.

The XPS spectra of deposited layers are shown in Figure 3. The observed peaks (Figure 3a) originate from Al2p, Al2s, O1s and C1s core electrons. In addition, O (KKL) Auger signals were easily identified. C1s contribution is believed due to the contamination during the preparation of the samples respectively from rest oil in the pump. In a higher magnification (Figure 3b) only one peak, which corresponds to Al_2O_3, can be seen. Due to size effects and the effective wrap-up of the interior Al wire an additional Al peak was not observed.

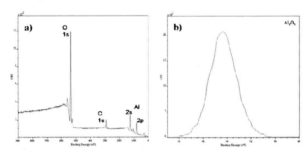

Figure 3. XPS analysis of coated glass surfaces (a) low magnification and (b) high magnification

Phase contrast micrographs (Figure 4a and 4b) show different behavior of the neuronal and glial cell cultures on non-coated and coated glass slides, respectively. It is clearly seen that neurite formation increases on 1D Al/Al_2O_3 nanostructures deposited glass slides compared to the non-coated glass slides. Fluorescence microscope images given in Figure 4c and 4d indicate similar observations. DRG neurons form a much denser neurite network on 1D Al/Al_2O_3 nanostructures, as compared with control substrates (non-coated glass). Optical and fluorescence microscope analysis indicate the potential of 1D Al/Al_2O_3 nanostructures to support the formation of communicative neural networks in three dimensions. Future investigations are needed to explore the interaction of neurons with such 1D Al/Al_2O_3 nanostructures in detail.

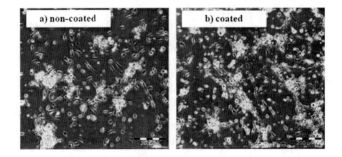

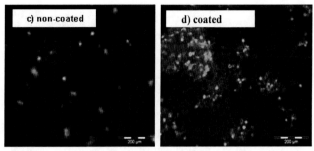

Figure 4. Phase contrast optical microscope images of neuronal and glial cell cultures on (a) non-coated and (b) coated round shaped glass substrates. Fluorescence microscope images of neuronal and glial cell cultures on (c) non-coated and (d) coated round shaped glass substrates

CONCLUSION

This preliminary study on neurons shows that 1D Al/Al$_2$O$_3$ nanostructures provide an enhanced cellular adhesion and growth which can be interesting for various applications in the medical fields as well as biosciences. In addition such 1D Al/Al$_2$O$_3$ nanostructures can be used to study the cellular communication and neural signaling. In order to investigate the only topography effects, different densities of the 1D Al/Al$_2$O$_3$ nanostructures will be deposited while keeping the surface chemical composition identical.

ACKNOWLEDGEMENT

The authors thank Dr. Vladimir Zaporojtchenko for the analysis of surfaces by XPS and the DFG (Deutsche Forschungsgemeinschaft) in the framework of SP 1181 on the theme "Komposit-Materialien aus molekularen Precursoren über Tandem CVD und Sol-Gel Techniken" for the financial support.

REFERENCES
[1] G. A. Silva; Nat. Rev. Neurosci. 7 (2006)
[2] U. Schroeder, P. Sommerfeld, S. Ulrich, B. A. Sabel; J. Pharm. Sci. 87 (1998),1305
[3] F. Yang, R. Murugan, S. Ramakrishna, X. Wang, Y. X. Ma, S. Wang; Biomaterials 25, 10 (2004), 1891
[4] D. H. Kim, M. Abidian, D. C. Martin; J. Biomed. Mater. Res. A 71 (2004), 577
[5] J. S. Wegman, A. Dwarka, M. Holzer, W.-K. Lye, M. L. Reed, E. Herzog, T. N. Blalock; Mat. Res. Soc. Symp. Proc. 773 (2003)
[6] M. C. Lieber Y. Li, J. Xiang; Mater. Today 9, 10 (2006), 18
[7] T. E. McKnight, A. V. Melechko, G. D. Griffin, M. A. Guillorn, V. I. Merkulov, F. Serna, D. K. Hensley, M. J. Doktycz; Nanotechnology 14 (2003), 551
[8] W. Hällström, T. Martensson, C. Prinz, P. Gustavsson, L. Motelius, L. Samuelson, M. Kanje; Nanoletters 7, 10 (2007), 2960
[9] M. Veith, S. J. Kneip, S. Faber, E. Fritscher; Mater. Sci. Forum 269 (1998), 303
[10] M. Veith, S. Faber, H. Wolfganger, V. Huch; Chem. Ber. 129 (1996), 381
[11] U. Rauch, A. Hänsgen, C. Hagl, S. H. Cunz, K. H. Schäfer; Int. J. Colorectal Dis. 21 (206), 554
[12]M. Veith, E. Sow, U. Werner, C. Petersen, O. C. Aktas; Eur. J. Inorg. Chem. 33 (2008), 5181
[13] Y. Wu, P. J. Yang; J. Am. Chem. Soc.123, 13 (2001), 1365

BIOACTIVE GLASS-CERAMIC/MESOPOROUS SILICA COMPOSITE SCAFFOLDS FOR BONE GRAFTING AND DRUG RELEASE

Enrica Verné, Francesco Baino, Marta Miola, Giorgia Novajra, Renato Mortera, Barbara Onida, Chiara Vitale-Brovarone

Materials Science and Chemical Engineering Department, Politecnico di Torino, Corso Duca degli Abruzzi 24, Torino, Italy

ABSTRACT

This work is focused on the development of an implantable system coupling bone regeneration and drug release ability. Specifically, ordered mesoporous silica spheres (MCM-41) were loaded onto a bioactive glass-ceramic scaffold. The macroporous scaffolds were prepared using fluoroapatite containing glass-ceramic (Fa-GC) powders and polyethylene (PE) particles of two different sizes. Fa-GC was prepared by a traditional melting-quenching route and was ground and sieved to obtain powders below 30 μm which were mixed with PE particles and then pressed in order to obtain crack-free green samples. The "greens" were thermally treated to remove the organic phase and to sinter the Fa-GC powders. Composite systems were then prepared by soaking the scaffold into the MCM-41 synthesis batch. The samples were characterized through X-Ray Diffraction, morphological observations, density measurements, mechanical tests and *in vitro* tests. Ibuprofen was used as model drug for uptake and delivery analysis of the system. Composite scaffolds combining the Fa-GC bioactive behaviour with the drug uptake-delivery properties of MCM-41 silica micro/nanospheres were successfully obtained. In comparison with the MCM-41-free scaffold, the adsorption capacity of composite scaffolds was 10 times enhanced and the drug release behaviour was highly affected by MCM-41 mesophase inside the scaffold.

INTRODUCTION

During the last decade the progress in chemical, physical, material and biological sciences resulted in the possibility of bone tissue engineering, i.e. a biologically based method for repair and regeneration of natural tissue [1-2]. A key component in tissue engineering for bone regeneration is the scaffold that acts as a template for cells interactions and for the growth of bone-extracellular matrix to provide structural support to the newly formed tissue [3].

Many researchers have tried to define which properties are required for an optimal synthetic scaffold, in particular for bone tissue replacement [4]. First of all scaffolds need to be biocompatible; in addition, a three-dimensional (3-D) porous geometry, similar to that of cancellous bone, and the retention of mechanical properties after implantation are required for scaffolds in order to maintain a tissue space of prescribed size and shape for tissue formation. A pores content within 50-70 %vol. is necessary; in the case of ceramic scaffolds, a macroporosity of 100−500 μm is needed to promote bone cells attachment, and a microporosity of less than 10 μm should favour ions and liquids diffusion. Scaffold properties depend primarily on the nature of the biomaterial, on the fabrication process [5-6] and on the implant 3-D micro-architecture. Several class of materials, e.g. metals, ceramics, glasses, synthetic polymers, natural polymers and combinations of these materials to form composites, have been studied and proposed for scaffolding. Modern fabrication technologies allow to produce scaffolds of various size and shape and, in some cases, patient-designed grafts.

After a biomaterial implantation, a serious problem that might occur is the adverse inflammatory response of the body with the related complications, such as septicaemia. For this reason, anti-inflammatory drugs and antibiotics are given to the patient with a consequent increase of the healing time. For this reason, the development of systems coupling bone regeneration and drug delivery ability is a promising field of applied research. In fact, an implantable system able to deliver drug molecules presents a lot of advantages, such as the reduction of the total drug dosage introduced in the body, a precise drug setting in the implant zone and the tuning of the drug release kinetics.

In the last years, ordered mesoporous materials were proposed as candidates for drug release applications and since their discovery in early 1990s, the ordered mesoporous silicas attracted the interest of many scientists. Mesoporous systems possess a highly ordered mesoporous structure which can be tuned by changing the synthesis conditions, and can be prepared in different forms, such as thin films, monoliths or powders. One of the most interesting research field is the employment of mesoporous systems as host matrices for functional molecules, such as enzymes, dyes or drugs [7-8].

This research work is focused on the preparation and characterization of an implantable composite system, formed by a glass-ceramic scaffold of complex composition, containing fluoroapatite crystals, and MCM-41, in whose nanopores the drug was incorporated. At this purpose the scaffolds were used as carriers for MCM-41 micro-/nano-spheres and the composite system was loaded with the drug molecules. The procedure followed to prepare MCM-41 [9] allows to obtain spherical particles of sub-micronical size, with nanopores size of ~2 nm. The drug chosen for the uptake and release studies was ibuprofen, which is extensively adopted in literature for such studies [7].

MATERIALS AND METHODS

Scaffold basic material

A glass-ceramic containing fluoroapatite (FA) crystals and belonging to the SiO_2-CaO-Na_2O-K_2O-P_2O_5-MgO-CaF_2 system was used as starting material for scaffolding. The glass-ceramic (Fa-GC) had the following molar composition: 50% SiO_2, 18% CaO, 9% CaF_2, 7% Na_2O, 7% K_2O, 6% P_2O_5, 3% MgO. Fa-GC was produced by melting the raw products (SiO_2, $Ca_2P_2O_7$, $CaCO_3$, $(MgCO_3)_4 \cdot Mg(OH)_2 \cdot 5H_2O$, Na_2CO_3, K_2CO_3, CaF_2) in a platinum crucible at 1,550°C for 1 h and by then quenching the melt in water to obtain a frit that was subsequently ball milled and sieved to a final grain size below 106 μm.

Scaffolds fabrication

The scaffolds were prepared by mixing the Fa-GC powders with a thermally removable organic phase which acted as a pore former; specifically, polyethylene (PE) particles ranging within 90-150 μm (Wigley Fibers, Somerset, UK) were used for this purpose. Fa-GC powders and PE particles were carefully mixed together for 0.5 h in a plastic bottle by means of a roll shaker to obtain an effective mixing. In this research, the PE amount was set at 70% vol. to have the best compromise between high pores content and satisfactory mechanical strength.

Crack-free compacts of powders ("greens") were obtained through uniaxial dry pressing of the mixed powders (130 MPa for 10 s). The "green" bodies were shaped in form of bars (50 × 10 × 10 mm³) and thermally treated in air (800 °C for 3 h, heating rates set at 5 °C·min⁻¹) to remove the PE powders and to sinter the inorganic phase obtaining macroporous glass-ceramic scaffolds. The sintered bars were cut (Struers Accutom 5 apparatus) to obtain cubic-like macroporous scaffolds.

Characterization of Fa-GC-derived scaffolds

Wide-angle (2θ within 10-70°) XRD analysis (X'Pert Philips diffractometer with Bragg-Brentano camera geometry, Cu anode Kα radiation with wavelength λ = 1.5405 Å) was carried out on the scaffold reduced in powders to detect the presence of crystalline phases nucleated during the thermal treatment.

Scaffolds morphology and microstructure were evaluated by scanning electron microscopy (SEM, Philips 525 M) to assess the pores size and distribution in the prepared samples; the samples were silver coated before examination. The scaffolds were carefully polished by SiC grit papers to finally obtain $10 \times 10 \times 10$ mm^3 samples that underwent pores analysis and mechanical testing.

The porosity content P (%vol.) was assessed by geometrical weight-volume evaluations on five specimens for each series as $P = \left(1 - \dfrac{\rho_s}{\rho_0}\right) \cdot 100$, where ρ_s is the apparent density of the scaffold (mass/volume ratio) and ρ_0 is the density of non-porous Fa-GC.

The scaffolds underwent crushing tests (MTS System Corp. apparatus, cross-head speed set at 1.0 mm·min^{-1}) carried out on five specimens for each series. The failure compressive stress σ_c (MPa) was assessed as $\sigma_c = \dfrac{F}{A}$, where F (N) is the maximum load registered during the test and A (mm^2) is the cross-sectional area perpendicular to the load axis.

The bioactive properties of the scaffolds was investigated *in vitro* by soaking the samples for different time frames in 30 ml of acellular standard Simulated Body Fluid (SBF) prepared according to Kokubo's protocol [10]. The solution was replaced every 48 h to approximately simulate fluids circulation in the human body; the pH variations of the solution were daily monitored. The samples were analyzed by SEM and EDS (Philips Edax 9100).

Preparation of composites scaffolds

Mesoporous MCM-41 nanospheres were synthesised and precipitated directly inside the scaffolds in order to obtain a composite material. The MCM-41 synthesis solution was prepared modifying a standard recipe [9]; n-hexadecyltrimethylammonium bromide (C$_{16}$TMABr) was used as a surfactant template. The molar ratio of the reactants was: 1 TEOS : 0.3 C$_{16}$TMABr : 0.129 NH$_3$: 144 H$_2$O : 58 EtOH. C$_{16}$TMABr was dissolved in a batch of distilled water, NH$_3$ and EtOH; tetraethylorthosilicate (TEOS, [C$_2$H$_5$O]$_4$Si) was then added to the surfactant solution under continuous stirring for 10 min. Afterwards, the cubic scaffold was soaked for 12 min directly in the solution at pH ~9. Hence, the MCM-41 spheres of sub-micronical size precipitated onto the scaffold and inside its macropores. Finally, the impregnated scaffold were aged at 90 °C for 24 h in air and then calcined through a two steps process (1 h at 200 °C in nitrogen, 7 h at 450 °C in air).

Characterization of composite scaffolds

Mechanical compressive tests and in vitro tests by soaking in SBF were carried out on composite as previously described concerning as-done scaffolds.

Drug uptake and release

The ability of the composite scaffolds to load molecules of ibuprofen was tested. At this aim, at room temperature, 9 ml of ibuprofen in pentane solution (33 mg ml^{-1}, 0.160 M) was put into contact with the composite scaffold for 72 h without stirring. The drug uptake was evaluated through UV-visible spectrophotometry (Cary 500 Scan, wavelength $\lambda = 263$ nm, molar extinction coefficient $\varepsilon_\lambda = 320$ M^{-1}cm^{-1}) and was assessed in terms of ibuprofen concentration variation in the solution before and after the scaffold soaking. The drug release kinetics were evaluated *in vitro* by soaking the composite scaffolds in 30 ml of stirred SBF maintained at 37 °C. At different time frames, a small amount of SBF (1 ml) was picked-up and analysed through UV-visible spectrophotometry ($\lambda = 263$ nm, $\varepsilon_\lambda = 440$ M^{-1}cm^{-1}) to assess the amount of released ibuprofen.

RESULTS AND DISCUSSION

Figure 1a shows the XRD pattern of as-poured Fa-GC: the main reflections actually corresponding to fluoroapatite were detected, marked and indexed, demonstrating the glass-ceramic nature of the material.

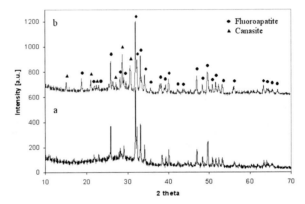

Figure 1. XRD analysis on (a) as-poured Fa-GC and (b) Fa-GC-derived scaffold.

Macroporous Fa-GC-derived scaffolds

The thermal treatment, carried out to produce the porous scaffolds, induced the partial crystallization of the residual amorphous phase into a new crystalline phase that was indexed as canasite ($K_3(Na_3Ca_5)Si_{12}O_{30}F_4 \cdot H_2O$), as shown in figure 1b.

The obtained scaffolds are characterized by open and interconnected macropores; for instance figure 2 show the presence of interconnected and homogeneously distributed macropores, which is a very important feature as a high degree of interconnection is crucial to attain a fast viability of the inner parts of the scaffold and to promote an effective bone in-growth in vivo.

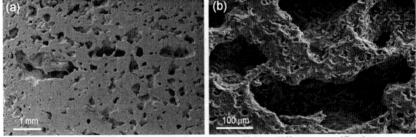

Figure 2. Cross-section of a Fa-GC-derived scaffold at different magnifications.

The SEM investigations showed that a high degree of sintering was achieved as it is possible to see in figure 2b. Inn addition, it should be noticed that the trabeculae are characterized by a dispersed microporosity (figure 2b), which is known to be useful for cell adhesion in vivo, and in the present work may also help the anchorage of MCM-41 spheres. Besides, the presence of micropores is essential for the flow of nutrients and waste products inside the scaffold in order to allow a proper environment for cells in the whole structure. The scaffold roughness (figure 2b) due to the presence of crystalline phases is also an important parameter because it was demonstrated that osteoblasts prefer a rough surface on which attach and also because coarseness can promote the anchorage of MCM-41 spheres on the scaffolds surface and inside its pores.

The total porosity, resulting from density measurement, is 50.1 ± 4.0 %vol.

The results of compressive tests performed on the prepared Fa-GC-derived scaffold were very satisfactory, because the samples showed a mechanical strength within 12-20 MPa, which is comparable with the strength of cancellous bone (2-20 MPa). Besides, the low standard deviation affecting the compressive strength values confirms the good repeatability of the fabrication method. The high strength is due to the peculiar morphology of Fa-GC-derived scaffolds in which the pores are separated by dense regions; this involves low interconnection degree of the pores. It should be noticed that, at present, the bioceramic scaffolds commonly proposed in literature for bone replacement, such as HA-based and Bioglass®-derived scaffolds, exhibit lower mechanical strength (below 1 MPa) than that of cancellous bone, and for this reason they are far from an actual clinical use. The features of the proposed Fa-GC-derived scaffolds make them very versatile grafts for bone replacement and they can be successfully proposed for the substitution of extensive bone portions also in load-bearing bone segments.

The composite scaffolds were tested in compression under the same conditions adopted for PE1-70 and PE3-50 samples. No significant differences, in terms of mechanical strength, were observed among the as-done and composite scaffolds of both the series. This is a very promising result, showing that MCM-41 synthesis batch does not degrade the scaffold structure.

In vitro bioactivity assessment

After soaking in SBF for 7 days, Fa-GC-derived scaffolds walls are completely covered by globe-shaped agglomerates of a newly formed phase, as shown in figure 3a.

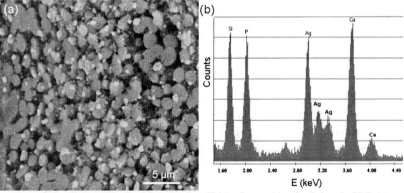

Figure 3. In vitro tests on Fa-GC-derived scaffolds after soaking for 7 days in SBF: (a) surface morphology and (b) compositional analysis.

The compositional analysis, reported in figure 3b, reveals that this phase is constituted by calcium and phosphorus, with Ca/P molar ratio of 1.65, which closely approaches the ratio of natural hydroxyapatite (1.67) in the natural bone. The peak of silicon (Si) is due to the silica-rich layer underlying the newly formed hydroxyapatite HA and whose formation is the basis of the bioactive process. The presence of a HA layer is expected to impart properties of biomimicry to the scaffolds and to promote cells adhesion. The variations of pH solution values were moderate: pH increased up to 7.55 after soaking for the first 48 h (reference value for SBF pH: 7.40). For this reason, no cytotoxic effect after *in vivo* implantation is foreseen to be induced by the material.

The *in vitro* bioactivity of both as-done scaffolds and of MCM-41-loaded ones was assessed by soaking them in SBF for 7 days and 1 month. No significant differences were observed between MCM-41-loaded scaffolds and as-done scaffolds during the bioactivity tests.

Drug uptake and release

The analysis of the composite scaffolds is mainly focused on two points: (i) to investigate if MCM-41-Fa-GC composite scaffold absorbs a greater amount of ibuprofen in comparison with non-loaded scaffolds and (ii) to investigate if MCM-41 inside Fa-GC scaffold plays a role in ibuprofen molecules release, in view of controlled drug delivery. For these purposes, a Fa-GC scaffold, used as reference (herein denoted as Scaff-ref), has been soaked in the MCM-41 synthesis solution without TEOS. The analysis time for ibuprofen uptake is 72 h. As it can be observed in table I, the amount of ibuprofen absorbed by MCM-41-Fa-GC is about 10 times greater than the Scaff-ref one.

Table I. Ibuprofen uptake on composite scaffolds

Scaffold	g IBU/g scaffold	% wt.
Scaff-ref	7.82	0.78
MCM-41-Fa-GC	80.01	8.00

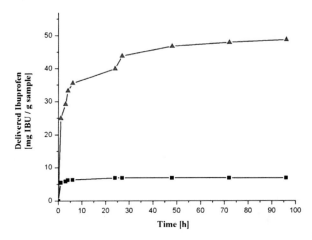

Figure 4. Ibuprofen release from Scaff-ref (squares). and MCM-41-Fa-GC (triangles).

The analysis time for ibuprofen release was 96 h. As it can be observed in figure 4, MCM-41-Fa-GC delivers a greater amount of drug in comparison with Scaff-ref because the latter one, during the uptake, absorbed a smaller ibuprofen amount; these results are consistent with the uptake ones. This demonstrated that ibuprofen release does occur from the mesoporous phase. It is worth underlining that the choice of the mesophase was strongly related to the adopted drug molecules (ibuprofen in the present work). In fact, it is crucial to obtain host-guest interactions among the mesophase nanopores and the drug molecules in order to promote a good drug uptake. For this reason, the pores size of prepared MCM-41 spheres (within 1.9-2.0 nm at pH = 9.0) is comparable to ibuprofen molecule size (molecule length ~1.0 nm).

CONCLUSION

3-D porous scaffolds for bone tissue replacement were produced by using bioactive glass-ceramic powders as basic material and polyethylene particles that were removed by thermal treatment. After sintering, porous samples having a porosity up to 50% vol. were produced. The fabrication method chosen for scaffolds production led to highly reproducible samples in terms of pores content and mechanical strength (up to 20 MPa). The 3-D network of interconnected macropores ranging within 100-300 μm can to promote in vivo blood vessels access and cells migration into the scaffolds. In addition, a diffused microporosity, which is known to promote cells adhesion, was observed in the prepared scaffolds. In addition, the scaffolds exhibited an *in vitro* highly bioactive behaviour, as a thick hydroxyapatite layer was formed on samples surface after soaking in simulated body fluids. The scaffolds were used as carriers for MCM-41 spheres and then impregnated with ibuprofen, which later on will be delivered by the system. Neither MCM-41 synthesis process nor the presence of MCM-41 spheres, anchored to scaffolds walls, affect the scaffold bioactivity or degrade its structure. Furthermore, MCM-41-loaded scaffolds can absorb and deliver a greater amount of ibuprofen in comparison with non-loaded ones, assessing the key role played by the mesophase. Therefore, the prepared composite scaffolds can be proposed for bone replacement and in situ drug delivery applications.

REFERENCES

[1]P.X. Ma, Scaffolds for tissue fabrication, *Mater. Today*, **7**, 30-40 (2004).
[2]L.L. Hench and J.M. Polak, Third-generation biomedical materials, *Science,* **295**, 1014-7 (2002).
[3]V. Karageorgiou and D. Kaplan, Porosity of 3D biomaterial scaffolds and osteogenesis, *Biomaterials,* **26**, 5474-91 (2005).
[4]D.W. Hutmacher, Scaffolds in tissue engineering bone and cartilage, *Biomaterials,* **21**, 2529–43 (2000).
[5]C. Vitale-Brovarone, S. Di Nunzio, O. Bretcanu and E. Verné, Macroporous glass-ceramic materials with bioactive properties, *J. Mater Sci.: Mater. Med.*, **15**, 209-17 (2004).
[6]C. Vitale-Brovarone, F. Baino and E. Verné, High strength bioactive glass-ceramic scaffolds for bone regeneration, *J. Mater Sci.: Mater. Med.*, **20**, 643-53 (2009).
[7]A. Ramila, B. Munoz, J. Perez-Pariente and M. Vallet-Regí, Mesoporous MCM-41 as drug host system, *J. Sol-Gel Sci. Tecn.*, **26**, 1199-202 (2003).
[8]P. Horcajada, A. Ramila, K. Boulahya, J. Gonzalez-Calbet and M. Vallet-Regi, Bioactivity in ordered mesoporous materials, *Solid State Sci.,* **6**, 1295-300 (2004).
[9]M. Grun, K.K. Unger, A. Matsumoto and K. Tsutsumi, Novel pathways for the preparation of mesoporous MCM-41 materials: control of porosity and morphology, *Microp. Mesop. Mater.*, **27**, 207-16 (1999).
[10]T. Kokubo and H. Takadama, How useful is SBF in predicting in vivo bone bioactivity?, *Biomaterials,* **27**, 2907-15 (2006).

COMPARISON OF OXIDE AND NITRIDE THIN FILMS
– ELECTROCHEMICAL IMPEDANCE MEASUREMENTS AND MATERIALS PROPERTIES

Y. Liu, C. Qu, R.E. Miller, D.D. Edwards, J.H. Fan, P .Li, E. Pierce, A. Geleil, G. Wynick, and X. W. Wang*
Alfred University
Alfred, NY 14802
USA

ABSTRACT

Oxide and nitride films are being examined for anti-corrosion applications. In this study, aluminum oxide, cerium oxide, and silicon nitride thin films were fabricated via two different techniques. The thickness range of such films is between 90 nm and 3 micrometers. Electrochemical Impedance Spectroscopy (EIS) measurements were utilized to evaluate various films on carbon steel substrates, including Open Circuit Potential (OCP), polarization and impedance measurements. Based on the EIS results obtained, a silicon nitride film had the highest impedance modulus and the lowest corrosion current among the films tested. The materials properties were measured via X-ray diffraction (XRD), Fourier Transform Infrared Spectroscopy (FTIR), Scanning Electron Microscopy (SEM), Energy-dispersive Spectroscopy (EDS), and X-ray Photoelectron Spectroscopy (XPS).

INTRODUCTION

When two dissimilar metals are brought together, an electrical potential is produced. In the presence of a salt solution, an electrochemical cell, which can facilitate corrosion of the metal, is formed. In some of the new vehicle designs, aluminum and/or magnesium alloys are considered as light weight engine materials,[1] which in turn will be in contact with carbon steel fasteners such as bolts, nuts and washers. To prevent such corrosion, various coating schemes have been considered, including zinc coatings, organic coatings, and oxide-polymer nanocomposite coatings.[2] Various oxide and nitride coatings have been utilized as barrier layers. As an example, silicon nitride thin films are currently employed as etch-stoppers in semiconductor fabrication processes and are being considered as environmental barrier coating for gas turbine applications.[3] As another example, thick aluminum oxide coatings are being utilized as protective, insulating, and/or anti-corrosive layers (on engine blades). This suggests that there is a potential for these coatings as protective barriers for carbon steels in certain corrosive environments. To explore the feasibity of ceramic coatings in anti-corrosion applications, this investigation focuses on the electrochemical impedance behavior of the coatings. Cerium oxide, aluminum oxide and silicon nitride films were deposited on carbon steel using electron beam evaporation technique or plasma enhanced chemical vapor deposition (PECVD) technique, and were evaluated via electrochemical impedance spectroscopy (EIS) for their effectiveness as protective coatings.

EXPERIMENTS

As-received carbon steel substrates were ground/polished with silicon carbide sanding papers, using 1,200-grit paper during the final step. Thereafter, the substrates were polished with 9 micron diamond suspension on an UltraPol Polishing cloth, followed by a 3 micron diamond suspension on a Trident polishing cloth, and finally by MasterPrep alumina suspension on Microcloth (Buehler, Lake Bluff, IL). After the last polishing step, the substrates were cleaned in the ultrasonic cleaner with a dilute solution of Blue Gold industrial cleaner (Modern Chemical, Inc., Jacksonville, Arkansas.) The substrates were subsequently cleaned with isopropanol (IPA), and then force-air dried.

* Corresponding Author: fwangx@alfred.edu

Cerium oxide and aluminum oxide thin films were deposited via electron beam vapor deposition techniques with or without oxygen ion assistances.[4] In Figure 1, an illustration is provided for the vapor deposition technique with the ion assistance. There are three assemblies: an electron beam assembly located near the lower-center position (Telemark TT-36 power supply/controller, 6 KW, 4 pocket crucible), an oxygen ion source assembly located at the center-left corner (Veeco grid-less Mark I ion source), and a substrate assembly located above both the electron beam assembly and the ion source assembly. The substrate assembly has two types of rotations: main rotating platform and satellites (double rotation planetary fixture). The target material for the evaporation, such as cerium oxide or aluminum oxide, is contained in a crucible.[5] Voltage range for the electron beam is between 7.4 and 8.0 KV; and current range is between 0.07 and 0.10 A. When the electron beam strikes the target material, vapor is produced, travels towards substrates (1A, 1B, 2A, and/or 2B), and condenses onto the substrates. Oxygen flow can be supplied to the chamber via two channels, with the total oxygen flow rate being 20 - 30 sccm. One of the channels is utilized for the oxygen ion source assembly. When the oxygen ion-beam (with 100 - 170 V and 1.35 - 1.77 A) strikes the substrate 2A and 2B surfaces, it has two purposes: "milling" and oxidation. That is, the function of the oxygen ion beam is to "mill" the recently coated oxide layer, and to facilitate full oxidation. During the deposition, the substrate temperature is approximately 190 °C. The chamber pressure is 0.5 micro-Torr or less before the deposition; and approximately 100 micro-Torr during the deposition. The film deposition rate is monitored by a quartz oscillator sensing device. The largest substrate size is 20 cm. Substrate materials include silicon wafers, soda lime glass and 1050 carbon steel.

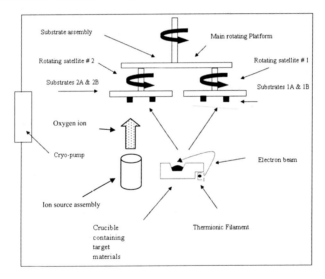

Figure 1. Illustration of electron beam vapor deposition technique with ion assistance

Silicon nitride films were fabricated via plasma enhanced chemical vapor deposition (PECVD) technique, illustrated in Figure 2.[6] A radio frequency (RF) power generator is connected to a top electrode and a bottom electrode which is also a substrate holder. Precursor gases, such as silane, ammonia and nitrogen, are fed into a vapor reaction chamber at flow rates of 130, 300, and 1,300 sccm respectively. These precursor gases are distributed via a shower head assembly. During deposition, the pressure in the chamber is approximately 950 milli-Torr, and the temperatures of the substrate holder and the shower head assembly are approximately 275 °C and 100 °C respectively. The RF frequency is approximately 13.56 MHz, and the power is approximately 250 W. The deposition time varies from 400 to 1,200 seconds. The film thickness is between 0.8 and 3 micrometers.

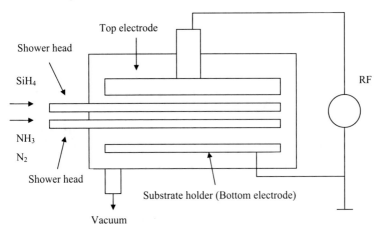

Figure 2. Illustration of PECVD setup

Morphology and topology of as-deposited thin film were characterized with an FEI Co. Quanta 200F environmental scanning electron microscope (ESEM). An acceleration voltage of 10-30 KV was utilized in high vacuum imaging mode. Microstructural characterization was completed under secondary electron imaging (SE). The presence of secondary phases was investigated under backscatter electron (BSE) imaging, in conjunction with an EDS (Energy Dispersive X-ray Spectroscopy) for elemental analysis. X-ray diffraction (XRD) analyses were conducted with a SIEMENS D-500 diffractometer. The 2-theta angles were between $10°$ and $100°$ with a scanning step of $0.04°$ and dwell time of 9 seconds. FTIR characterization of the samples was performed on NEXUS 670 in a diffuse reflectance mode, with an MCT.

Electrochemical Impedance Spectroscopy (EIS) measurements were carried out via Solartron potentiostat/galvanostat and frequency response analyze.[9] In order to prescreen the coated samples, an open circuit potential (OCP) measurement was first conducted. Then, a polarization measurement was carried out. Finally, an electrochemical impedance measurement was performed. In the EIS measurement setup, there were three electrodes: a working electrode (WE), a counter electrode (CE), and a reference electrode (RE). The RE was the Saturated Calomel Electrode (SCE).[11] The CE was the meshed platinum, with the size of the platinum being 1''x 1''. The WE was either an as-received carbon steel substrate or a coated steel substrate. In polarization measurements, the scanning rate was 1 mV/sec. The weight percentage of the sodium chloride in the distilled water solution was 5%.[11]

After EIS testing, thin film samples were analyzed using X-ray photoelectron spectroscopy (XPS) with a PHI Quantera SXM using monochromated Al Kα radiation.The beam size of the x-ray beam was 200 µm in diameter, with a beam power of 50 W and an operating potential of 15 kV. Areas under the resulting peaks (for each electron state), between each sputtering period, were evaluated using iterated Shirley background fitting. The resulting peak areas were scaled in accordance with standard sensitivity factors.[7] Atomic concentrations, prior to each period of sputtering, were calculated from the scaled peak areas. An atomic concentration profile was provided as a function of the sputtering time.[8] As illustrated in Figure 3, for a thin film sample, there were three areas studied via XPS. Area A was not exposed to sodium chloride solution, and there was no corrosion spot. Area B was exposed to sodium chloride solution, and there was no visible corrosion spot. Area C was also exposed to sodium chloride solution, and there was a visible corrosion spot.

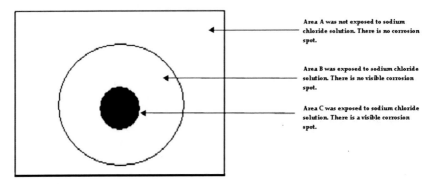

Area A was not exposed to sodium chloride solution. There is no corrosion spot.

Area B was exposed to sodium chloride solution. There is no visible corrosion spot.

Area C was exposed to sodium chloride solution. There is a visible corrosion spot.

Figure 3. Three areas of a thin film sample

RESULTS

For cerium oxide, the deposition rate is between 0.5 Å/s and 1.5 Å/s. The film thickness is between 140 nm and one micrometer. In Figures 4 (a) and (b), top views of SEM images (10K and 100K magnifications) are provided for a cerium oxide thin film on a steel substrate, fabricated with the electron beam evaporation technique at a substrate temperature of approximately 190 °C. The micrograph shows a granular microstructure with features on the order of 50 nm and notable porosity. Based on a cross-sectional view (not shown), the thickness of this film is approximately one micrometer. Figure 5 is an EDS spectrum for a 140 nm cerium oxide film, in which both cerium and oxygen are revealed. In addition, iron, manganese and silicon are revealed, which are from the 1050 carbon steel substrate. An XRD diffraction pattern (not shown) indicates the presence of a cerium dioxide phase in the film (ICDD 01-073-9516).[4] In Figure 6, impedance modulus values at frequency of approximately 0.1 Hz are plotted as a function of exposure time in NaCl solution. There are three samples. The first sample is uncoated 1050 carbon steel, the second sample is a one-micrometer cerium oxide film coated on 1050 steel via electron beam technique, and the third sample is a one-micrometer cerium oxide film coated on 1050 steel via ion assisted electron beam technique. The impedance modulus values of cerium oxide coated samples are slightly higher than that of the uncoated sample. In Figure 7, an XPS depth profile is provided for a cerium oxide film fabricated via electron beam technique (in Area A illustrated in Figure 3). The interfacial region between cerium oxide film and the steel substrate is visible. However, another XPS depth profile (not shown here) for Area C indicates that the interfacial region between cerium oxide film and the steel substrate is no longer in existence after the area has been exposed to NaCl solution for a week or so.

Figure 4 (a). Top view of a cerium oxide thin film on steel with 10K magnification

Figure 4 (b). Top view of a cerium oxide thin film with 100K magnification

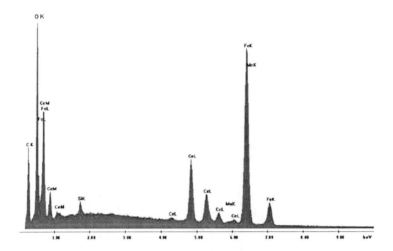

Figure 5. EDS spectrum of cerium oxide thin film on silicon substrate

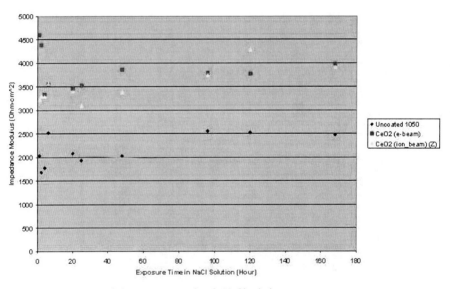

Figure 6. Impedance modulus vs. exposure time in NaCl solution

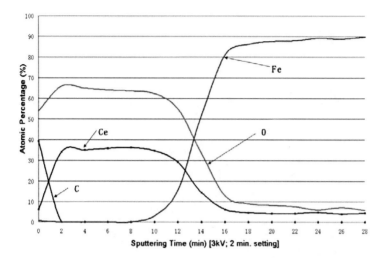

Figure 7. XPS depth profile of cerium oxide thin film on 1050 carbon steel substrate

Aluminum oxide thin films were deposited via electron beam evaporation technique illustrated in Figure 1. The substrate temperature was approximately 190 °C. The deposition rate is between 0.1Å/s and 1.0 Å/s. The film thickness is between 90 nm and 700 nm. Top views of SEM images are illustrated in Figures 8 (a) and (b), with magnifications of 10K and 100K respectively, for a film on a steel substrate. There are some pores in the aluminum oxide film. The pore distribution is less visible than that of the cerium oxide film illustrated in Figures 4 (a) and (b). In Figure 9, an EDS spectrum is provided for an aluminum oxide film coated on the steel. Besides the elements due to the coating such as aluminum and oxygen, elements in the substrate are also revealed, including iron, calcium, magnesium, sulfur and manganese. In Figure 10, the open circuit potential (OCP) is plotted as a function of time. The middle-blue curve is for a typical aluminum oxide thin film coated on 1050 carbon steel, and the bottom-green curve is for uncoated 1050 carbon steel. In general, the OCP value of the aluminum oxide film is higher than that of the uncoated carbon steel sample. In Figure 11, three polarization curves are provided, with the green curve for uncoated 1050 carbon steel and the blue curve for aluminum oxide coated 1050 steel. In Figure 12, impedance modulus is plotted as a function of the frequency. The middle-blue curve is for an aluminum oxide coating on the steel, and the bottom-green curve is for the uncoated steel. At a given low frequency, say 0.1 Hz, the impedance modulus for the aluminum oxide coated sample is higher than that of the uncoated sample by an order of the magnitude. In Figure 13, an XPS depth profile of an aluminum oxide thin film on the steel substrate is shown, for Area A illustrated in Figure 3. The interfacial region between the coating and the substrate is visible. However, a depth profile (not shown here) for Area C illustrated in Figure 3 does not show any interfacial region. For Area B illustrated in Figure 3, a depth profile (not shown here) reveals a "smeared" interfacial region.

Figure 8 (a). Top view of aluminum oxide film on steel with 10K magnification

Figure 8 (b). Top view of aluminum oxide film with 100K magnification

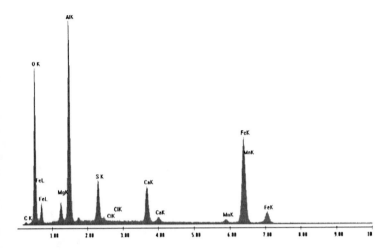

Figure 9. EDS spectrum of an aluminum oxide coating on steel

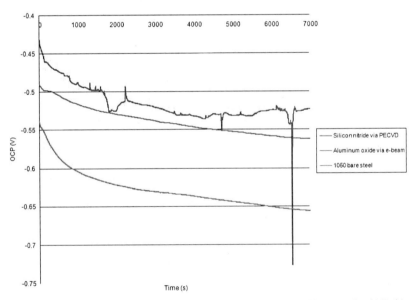

Figure 10. Open circuit potential vs. time. Top-red curve is for a silicon nitride on steel, middle-blue curve is for an aluminum oxide thin film on steel, and the bottom-green curve is for uncoated steel.

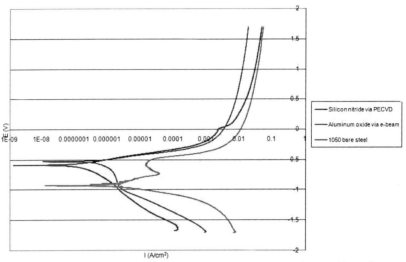

Figure 11. Polarization curves for uncoated steel substrate (green), aluminum oxide coating on steel (blue) and silicon nitride coating on steel (red)

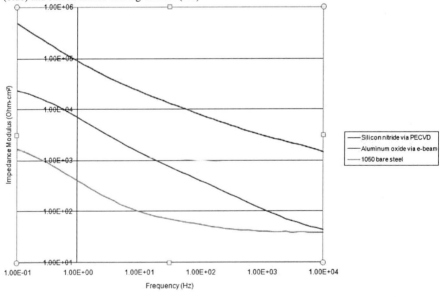

Figure 12. Impedance modulus vs. frequency: Top-red curve is for silicon nitride coating on steel, middle-blue curve is for aluminum oxide coating on steel and bottom-green curve is for control.

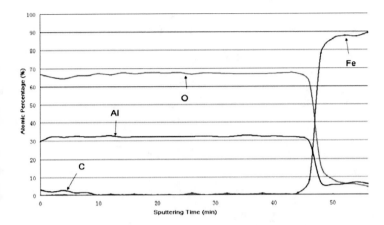

Figure 13. XPS depth profile of Aluminum oxide coating on steel, for Area A in Figure 3

Silicon nitride films were analyzed via SEM/EDS, XRD and FTIR measurements. Secondary electron (SE) SEM images taken normal to the film surface can be found in Figures 14 (a) and (b), for a film on a steel substrate, with magnifications of 10K and 100K respectively. The coating appears to be nearly continuous. The EDS spectrum of a silicon nitride film is presented in Figure 15, in which silicon, nitrogen and oxygen are shown. The signal of carbon originates from the conductive coating used in preparing this sample. XRD pattern (not shown) reveals the presence of silicon nitride phase (ICDD 00-009-0245) in the film. In Figure 16, an FTIR spectrum for a silicon nitride film is illustrated, for a film with thickness being approximately 0.85 micrometers. There are several vibrational modes including Si-H, N-H and Si-N bonds. The signature of Si-N stretching mode is around 850 cm^{-1}, the signature of Si-H stretching mode is around 2,200 cm^{-1}, and the signature of N-H is around 3,200 cm^{-1}. The OCP curve for the silicon nitride film is illustrated in Figure 10, which is relatively unstable in comparison with that of the aluminum oxide film. The polarization curve for silicon nitride coated steel is illustrated in Figure 11. The corrosion current density of the silicon nitride sample is lower than that of the aluminum oxide sample (Figure 11), and that of the cerium oxide sample (not shown). The impedance modulus of silicon nitride sample is plotted as a function of the frequency in Figure 12. At 0.1 Hz, the impedance value of the silicon nitride sample is approximately one order of magnitude higher than that of the aluminum oxide sample. Figure 17 is XPS depth profile for silicon nitride film on steel (Area A in Figure 3). The interfacial region between the silicon nitride film and the steel substrate is clearly visible. Figure 18 is depth profile for Area B (in Figure 3). After contacting NaCl solution, the silicon and nitrogen curves in Figure 18 are changed slightly, in comparison with that in Figure 17. Figure 19 is depth profile for Area C (in Figure 3), and shows evidence of corrosion.

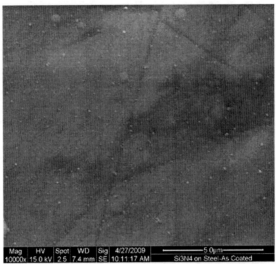

Figure 14 (a). Top view of silicon nitride on steel with 10K magnification

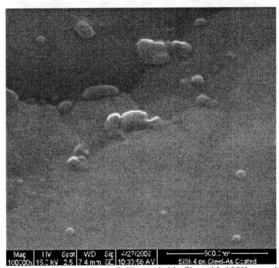

Figure 14 (b). Top view of silicon nitride film with 100K magnification

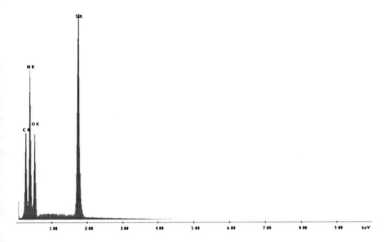

Figure 15. EDS Spectrum of Silicon Nitride

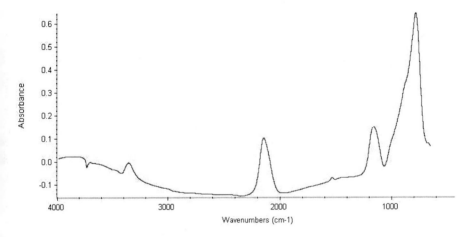

Figure 16. FTIR spectrum for silicon nitride film in Area A of Figure 3

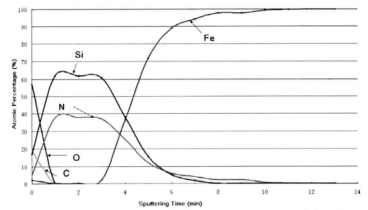

Figure 17. XPS depth profile of silicon nitride film on steel (Area A of Figure 3)

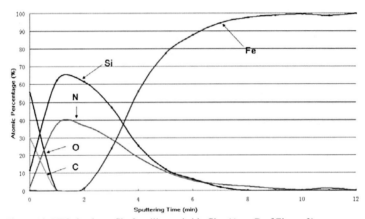

Figure 18. XPS depth profile for silicon nitride film (Area B of Figure 3)

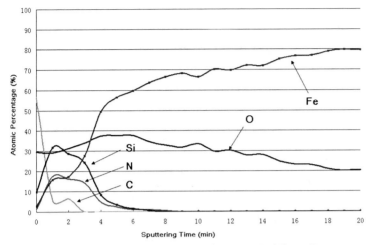

Figure 19. XPS depth profile for Silicon nitride film (Area C of Figure 3)

CONCLUSION AND DISCUSSION

Cerium oxide and aluminum oxide films were fabricated via an electron beam evaporation technique, and silicon nitride films were fabricated via PECVD. Based on the top view, the silicon nitride coating almost has no pores, the aluminum oxide has some pore distribution, and cerium oxide coating has a lot of pores. Based on EIS measurements, the silicon nitride coating appears to have the best anti-corrosion properties. That is, the silicon nitride film material has the lowest corrosion current density and the highest impedance modulus value at $f = 0.1$ Hz. As far as the anti-corrosion property is concerned, the aluminum oxide coating is worse than the silicon nitride coating, but better than the cerium oxide coating. In order to understand the connections between the film topology/morphology and the corrosion current flow, more detailed analyses are being carried out currently. In addition, there are two issues to be resolved. The first issue is the relationship between the film thickness and the anti-corrosion performance. Presently, we are systematically using EIS measurements to establish the connections between film thickness and impedance modulus, providing that all the films are fabricated under the same conditions besides the deposition time. The second issue is related to the stoichiometric ratios of Ce:O, Al:O and Si:N:O in the films. Presently, we are utilizing EDS, WDS and XPS to systematically study the ratios, which may be related to the corrosion rates.

ACKNOWLEDGMENT

Help from following people is appreciated: Dr. W. A. Schulze, R. J. Lewis, A. Ezell, V. Dansoh, F. T. Gertz, Y. D. Liu, Dr. G. J. McGowan, Dr. A. Eklund, Dr. A. Meier, Dr. W. Mason, and S. Zdzieszynski.

DISCLAIMER

This material is based upon work supported by the Department of Energy National Energy Technology Laboratory under Award Number DE-FC26-02OR22910. This report was prepared as an account of work sponsored by an agency of the United States Government. Neither the United States Government nor any agency thereof, nor any of their employees, makes any warranty, express or implied, or assumes any legal liability or responsibility for the accuracy, completeness, or usefulness of any information, apparatus, product, or process disclosed, or represents that its use would not infringe privately owned rights. Reference herein to any specific commercial product, process, or service by trade name, trademark, manufacturer, or otherwise does not necessarily constitute or imply its endorsement, recommendation, or favoring by the United States Government or any agency thereof. The views and opinions of authors expressed herein do not necessarily state or reflect those of the United States Government or any agency thereof. The partial financial support from US CAR program is acknowledged.

REFERENCES
[1]See, for example, E.A. Nyberg, A.A. Luo, K. Sadayappan, and W. Shi, "Magnesium for Future Autos," *Adv. Mater. Proc.*, **166** [10] 35-37 (2008).
[2]See, for example, Y. Wang, S. Lim, J.L. Luo, and Z.H. Xu, "Tribological and Corrosion Behaviors of Al_2O_3/polymer Nanocomposite Coatings," *Wear*, **260** [9-10] 976-983 (2006).
[3]S. Ueno, D. Jayaseelan, N. Kondo, T. Ohji, S. Kanzaki and H. Lin, "Development of EBC for Silicon Nitride," Key Engineering Materials, **287**, 449-456, 2005.
[4]V. Dansoh, F. Gertz, J. Gump, A. Johnson, J. Jung, M. Klingensmith, Y. Liu, Y. D. Liu, J. Oxaal, C. Wang, G. Wynick, D. Edwards, J. Fan, X. W. Wang, P. Bush, and A. Fuchser, "Cerium Oxide Thin Films via Ion Assisted Electron Beam Deposition," Ceramic Engineering and Science Proceedings, **29**, 87-98 (2008).
[5]The purity of a target material is between 99.9% and 99.99%.
[6]Trikon/Aviza Planar FxP PECVD system, with the deposition chamber size being approximately 0.6 m X 0.6 m X 0.3 m
[7]Handbook of X-ray Photelectron Spectroscopy: A Reference Book of Standard Spectra for Identification and Interpretation of XPS Data, edited by J.F. Moulder, W.F. Stickle, P.E. Sobol, and K.D. Bomben. Physical Electronics, Eden Prairie, Minnesota, 1995.
[8]Due to different sputtering rate for different material, the sputtering time is a relative measurement for the film thickness.
[9]Solartron potentiostat/galvanostat 1287 and Solartron 1260 frequency response analyzer (FRA)
[10]Fisher Scientific, Inc., catalog number 13-620-79
[11]The pH value of the solution was approximately 7.3, the temperature in the lab was approximately 20 °C, and the humidity in the lab was approximately 78%.

SYNTHESIS OF PbTe NANOWIRES WITH ENHANCED SEEBECK COEFFICIENT

Wenwen Zhou[1], Hao Cheng[1], Aidong Li[2], Huey Hoon Hng[1], Jan Ma[1], Qingyu Yan[1,2]
[1]School of Materials Science and Engineering, Nanyang Technological University, Singapore
[2]Materials Science and Engineering Department, Nanjing University, Nanjing, China

ABSTRACT:
    A simple synthesis protocol for PbTe nanowires (NWs) preparation with controlled size, e.g. diameter 10~30 nm and length 500~3000 nm, and enhanced seebeck coefficient was demonstrated. The 1D growth of PbTe NWs was directed due to the surface bonds between Pb and sucrose, which formed the soft stacking template due to the π-π electron interaction. The size and morphology of the NWs were controlled with respect to the reaction temperatures and heating rates, e.g. by injecting the precursors to the solvent directly at 180 °C or by heating the precursors gradually from room temperature to 180 °C. Very high p-type seebeck coefficients >470 μV/K at T = 375~425 K were obtained in the film samples made from PbTe NWs after hydrazine washing and annealing at 300 °C in vacuum. The high seebeck coefficient value was suspected to be related to quantum size confinement or due to phonon drag scattering of the charge carriers. These NWs may give promising potentials for architecture of thermoelectric module applications.

INTRODUCTION:
    Thermoelectric modules have recently attracted great interests for applications in power generation from waste heat and solid state cooling[1]. The performance of the thermoelectric modules depends on the dimensionless figure-of-merit (ZT), defines[2] as $ZT = S^2 \sigma T / k$, where S is the seebeck coefficient, σ the electrical conductivity, k the thermal conductivity and T the absolute temperature. Many exciting progress has been developed in bulk materials, e.g. silver antimony lead telluride (LAST)[3] and bismuth antimony telluride[4]. Theories have predicted that decreasing the size of the thermoelectric materials can increase ZT beyond the bulk values[5, 6], e.g. superlattice thin films[7, 8] and nanowires[9, 10]. Nanowire/nanorods are of particular interest because they are suitable for building device circuit, which allows the study of thermal and electrical properties of individual nanostructure through contact formation.
    Normally, different thermoelectric materials show peak ZT values in different temperature ranges. PbTe is of particular interest because it has the highest ZT of any single-phase bulk solid in the temperature[11] range of 400-700 K. Many progresses have been developed for the synthesis of PbTe nanoparticles, e.g by solvothermal process[12-16]. However, the synthesis of PbTe based nanowires/nanorods still largely depends on electrochemical deposition into hard templates[17, 18], which gives rise to polycrystalline products and requires a tedious process to remove the hard template for further device integration. Limited success has been demonstrated in soft-templating approaches, which utilize molecular agents to direct the growth of single crystal nanowires to achieve high ZT values and easy device architecture.
    Here, we introduce a simple solvothermal polyol process to prepare p-type PbTe nanowires with controlled size, e.g. with diameter varying from 10 nm to 40 nm and length varying from 600 nm to 3 μm. Their promising thermoelectric properties are also presented. Surfactant sucrose was used to direct the 1D growth of PbTe nanowire as well as the reducing assist of precursors. Without sucrose, the reaction time was prolonged and only large cubic particles were formed. The size of the nanowires was highly dependent on the reaction temperature and reaction time. The film samples made from the PbTe nanowires yielded high seebeck coefficients >470 μV/K in the temperature range of 375-425 K.

EXPERIMENT:

PbTe NWs synthesis

Lead acetate trihydrate (99%+), orthortelluric acid (H_6TeO_6 99.99%), 1-pentanediol, sucrose, trioctylphosphine oxide and hydrazine monohydrate ($N_2H_4 \cdot H_2O$) were obtained from Aldrich and used without further purification. In a typical synthesis, 23 mg orthortelluric acid (0.1 mmol) and 114 mg lead acetate trihydrate were dissolved into 1 mL pentanediol in two separate vials at T = 40 °C in ambient, respectively. In a separate three-neck round flask, 100 mg sucrose were added into 20 mL pentanediol and heated to 180 °C for 5 minutes under Ar flow. The orthortelluric acid and lead acetate trihydrate in pentanediol solution were then mixed and the mixture were injected into three-neck round flask under moderate magnetic stirring. Upon injection, black precipitates were observed within 30 seconds. The reaction was stopped after 1-2 minutes by quenching the round flask in an ice water bath. The black precipitates were cleaned by repeated washing with ethanol and centrifuging. The NWs were then dispersed in ethanol and stored in N_2 gas environment, which were used for SEM, TEM, EDAX, HRTEM and XRD characterization.

Materials characterization

The as-prepared PbTe NWs were deposited by drop casting onto 1×1 cm^2 silicon substrates for SEM and XRD measurements and onto carbon coated 200 mesh Cu grits for TEM, HRTEM and EDAX characterization. The X-ray diffractograms were obtained using a Scintag PAD-V diffractometer with a Cu Kα irradiation. The size and morphology of the lead telluride nanostructures was characterized using a field-emission SEM (JEOL JSM6335) operating at 5 kV. HRTEM and SAED measurements were carried out in a JEOL 2010 instrument operating at 200 kV.

Thermoelectric properties characterization

The as-prepared PbTe NWs in ethanol were centrifuged out and re-dispersed into the $N_2H_4 \cdot H_2O$: ethanol = 1:4 mixture to remove the capping surfactants. After 1 hour, the PbTe NWs were again centrifuged out and dispersed into the $N_2H_4 \cdot H_2O$: ethanol = 1:8 mixture. The solution was then used for film preparation onto glass substrate by using an air brush spray system. The thickness of the film was measured using a scanning probe station. For resistivity measurement, the PbTe NWs were spray deposited onto glass substrates with four pre-patterned Au lines. The resistivity was measured by four-probe technique on the films. The seebeck coefficients were measured on the film samples with an area of 3mm×2 cm using a ZEM-3 seebeck meter at pre-selected temperature range of 300~475 K.

RESULTS AND DISCUSSION:

Scanning electron microscopy (SEM) image in Fig. 1a reveals the PbTe nanowires obtained from samples prepared at 180 °C and quenched after 90 seconds. The nanowires are straight and have uniform lengths spanning 2-3 µm. Although the side wall of the wires is jagged, the nanowire width has a narrow distribution in the range of 15-25 nm (see Fig. 1a-b). Energy dispersive X-ray (EDX) analysis of the as-prepared sample revealed a Pb:Te ratio of close to 50:50, independent of the ratio between the precursors, lead acetate and telluric acid. Phase contrast atomic resolution TEM indicates that each nanowire is single crystal (see Fig. 1b). The long axis of the nanowires is parallel to $(\bar{2}42)$ planes and the short axis is parallel to (202) planes. X-ray diffraction (XRD) pattern from the as-prepared nanowires reveals the cubic crystal structure (see Fig. 1c), which is consistent with the HRTEM and SAED results. No impurity phases were detected in the XRD result.

The size and morphology of the PbTe nanowires are sensitive to the reaction temperature, the heating sequence as well as the surfactant template. Nanowires of 10-16 nm in width and 400-600 nm in length were obtained (see Fig. 1d) by injecting precursors into pentanediol heated at 210 °C and quenched after 60 seconds. We noted that the width of the PbTe nanowires is smaller than those reported to date. Nanowires with smaller width hold more promise because the increased phonon scattering decreases the thermal conductivity and the quantum size effect can enhance the seebeck coefficient[6]. However, dissolving lead acetate and telluric acid into pentanediol at room temperature and then heating the mixture gradually to 180-210 °C to allow the reaction to take place led to the formation of non-uniform shaped mixture (Fig. 2a-b). We also noted that sucrose is the key to direct 1D growth[19] in this particular reaction. Precipitation of PbTe under the same condition without sucrose led to the formation of cubes with an average size of ~100 nm (see Fig. 2c). The required reaction times for the cases without sucrose were prolonged to ~30 minutes as judged by the appearance of black precipitation. The reaction time for synthesis without sucrose is much longer as compared to with sucrose, which indicates that sucrose facilitates the reduction of the precursors as well as providing the soft templates to direct 1D growth.

Based on the above observation, we propose the phenomenological description that describes the key aspect of the PbTe nanowire formation. Although sucrose has been proven to assist 1D growth[19] by forming stacking template due to the π-π electron interaction as illustrated in scheme 1, the hydroxyl groups of sucrose interact only effectively with Pb but not Te. For example, replacing Pb with Pt in the same synthesis process led to the formation of highly agglomerated random-shaped PtTe crystals (See Fig 2d), which indicates the lack of passivation from sucrose on both Te and Pt. Thus, the precipitation sequence of Te and Pb is the key to allow anisotropic growth. During the process of gradually heating the mixture to 180-210 °C, Te precipitates much faster than that of Pb. This is due to the higher reduction potential of Te^{6+} as compared to that of Pb^{2+}, which forms large non-uniform shaped mixtures due to the absence of passivation by sucrose during the initial Te nucleation stage. Injecting the precursors into the solvent at high temperature accelerates the precipitation of Pb that mixes with Te. The Pb interacts effectively with sucrose to passivate the nanostructure and translate the templating effect from the sucrose to direct the 1D growth.

In order to assess the thermoelectric properties of the PbTe NWs, the NWs were sprayed onto glass substrates after hydrazine washing and then annealed at 300 °C in vacuum. The NW film is porous while the NWs are randomly aligned without noticeable coalescence (see Fig. 3a and b). Seebeck coefficient S measurement on the annealed NW film in the temperature range 315-475 K revealed p-type semiconductor behavior (see Fig. 3c). The S values are >470 µV/K at T = 375 K and 425 K. These values are much higher than those of pure bulk PbTe with various charge carrier concentrations. Although we speculate that the enhancement of S is due to quantum size effect induced modification of the density of state (DOS)[20] as the diameter of the NWs is 10~20 nm (which is close to the quantum size region of <10 nm), we cannot rule out the possibility of the phonon drag effect[10] caused by the increased scattering between phonon and charge carrier[21] as the length (1~3 µm) of the NWs are higher than the mean free path of phonons in PbTe. Further investigation is currently undergoing to verify this. The film exhibits a linear current-voltage (I-V) curve indicating the ohmic contact response. The resistivities of the annealed film are plotted at different temperatures as shown in Fig. 3d. Here, the variation of the film thickness contributes to the error of the resistivity values. The resistivity of the NWs films at room temperature is about 2 orders larger than that of bulk PbTe with a charge carrier concentration of 2×10^{18} cm^{-3} [reference 22]. This is possibly due to the un-optimized interconnection in the NW network. Here, we also noticed that the resistivity is highly dependent on the charge carrier concentration in the PbTe. The measurement of resistivity of individual NW is currently ongoing to check whether higher conductivity can be achieved in the PbTe NW.

In summary, we have demonstrated a simple polyol approach to synthesize PbTe NWs with tunable size. Sucrose passivation is the key to induce the one dimensional growth of PbTe. The size and morphology of the NWs are highly dependent on the reaction temperature and heating rates. Enhancement of the seebeck coefficient has been observed, which could possibly due to either quantum size effect or phonon drag effect or both. The mechanism is under further investigation. Although the resistivity measured in the NW films are still high due to the un-optimized interconnections, we expect the resistivity of individual NW to be lower. Further optimization of the thermoelectric properties of PbTe NWs based devices through interconnect alignment and doping may lead to energy conversion application of high efficiency.

ACKNOWLEDGEMENTS
The authors gratefully acknowledge the Start-Up Grant from NTU, AcRF Tier 1 RG 31/08 from MOE Singapore, and NSFC 10704035 Grant.

REFERENCES
[1]   H. J. Goldsmid, *Thermoelectric Refrigeration*, Plenum, New York **1964**.
[2]   B. C. Sales, *Science* **2002**, *295*, 1248.
[3]   K. F. Hsu, S. Loo, F. Guo, W. Chen, J. S. Dyck, C. Uher, T. Hogan, E. K. Polychroniadis, M. G. Kanatzidis, *Science* **2004**, *303*, 818.
[4]   B. Poudel, Q. Hao, Y. Ma, Y. C. Lan, A. Minnich, B. Yu, X. Yan, D. Z. Wang, A. Muto, D. Vashaee, X. Y. Chen, J. M. Liu, M. S. Dresselhaus, G. Chen, Z. Ren, *Science* **2008**, *320*, 634.
[5]   M. S. Dresselhaus, G. Dresselhaus, X. Sun, Z. Zhang, S. B. Cronin, T. Koga, J. Y. Ying, G. Chen, *Microscale Thermophysical Engineering* **1999**, *3*, 89.
[6]   M. S. Dresselhaus, G. Dresselhaus, X. Sun, Z. Zhang, S. B. Cronin, T. Koga, *Physics of the Solid State* **1999**, *41*, 679.
[7]   R. Venkatasubramanian, E. Siivola, T. Colpitts, B. O'Quinn, *Nature* **2001**, *413*, 597.
[8]   T. C. Harman, P. J. Taylor, M. P. Walsh, B. E. LaForge, *Science* **2002**, *297*, 2229.
[9]   A. I. Hochbaum, R. K. Chen, R. D. Delgado, W. J. Liang, E. C. Garnett, M. Najarian, A. Majumdar, P. D. Yang, *Nature* **2008**, *451*, 163.
[10]   A. I. Boukai, Y. Bunimovich, J. Tahir-Kheli, J. K. Yu, W. A. Goddard, J. R. Heath, *Nature* **2008**, *451*, 168.
[11]   D. M. Rowe, Ed. *CRC Handbook of Thermoelectrics*, CRC press, New York **1995**.
[12]   T. L. Mokari, M. J. Zhang, P. D. Yang, *Journal of the American Chemical Society* **2007**, *129*, 9864.
[13]   J. J. Urban, D. V. Talapin, E. V. Shevchenko, C. R. Kagan, C. B. Murray, *Nature Materials* **2007**, *6*, 115.
[14]   J. E. Murphy, M. C. Beard, A. G. Norman, S. P. Ahrenkiel, J. C. Johnson, P. R. Yu, O. I. Micic, R. J. Ellingson, A. J. Nozik, *Journal of the American Chemical Society* **2006**, *128*, 3241.
[15]   W. G. Lu, J. Y. Fang, K. L. Stokes, J. Lin, *Journal of the American Chemical Society* **2004**, *126*, 11798.
[16]   R. Kerner, O. Palchik, A. Gedanken, *Chemistry of Materials* **2001**, *13*, 1413.
[17]   W. F. Liu, W. L. Cai, L. Z. Yao, *Chemistry Letters* **2007**, *36*, 1362.
[18]   M. Sima, I. Enculescu, M. Sima, E. Vasile, *Journal of Optoelectronics and Advanced Materials* **2007**, *9*, 1551.
[19]   Q. Yan, M. S. Raghuveer, H. F. Li, B. Singh, T. Kim, M. Shima, A. Bose, G. Ramanath, *Advanced Materials* **2007**, *19*, 4358.
[20]   M. S. Dresselhaus, G. Chen, M. Y. Tang, R. G. Yang, H. Lee, D. Z. Wang, Z. F. Ren, J. P. Fleurial, P. Gogna, *Advanced Materials* **2007**, *19*, 1043.
[21]   J. P. Heremans, C. M. Thrush, D. T. Morelli, *Physical Review B* **2004**, *70*.

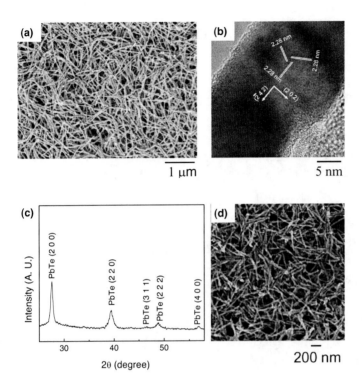

Figure 1. (a) A SEM image of PbTe nanowires prepared at 180 °C. Representing (b) TEM and (c) HRTEM images of a PbTe NW in the same batch as (a). (d) X-ray diffractogram of PbTe NWs as shown in (a)-(c). (e) SEM image of PbTe NWs prepared at 210 °C. (f) Typical TEM image of a PbTe NW in the same batch as (e). Inset in (f) is the corresponding SAED pattern.

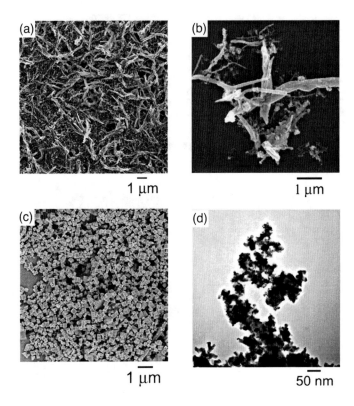

Figure 2. (a) and (b) both show the SEM images of the PbTe clusters formed by heating the precursors from room temperature gradually to 180 °C with a heating rate of 5 K/min. The structures are not uniform. (c) PbTe cubes precipitate in the synthesis without sucrose. (d) TEM micrograph of PtTe$_2$ agglomerated crystals synthesized at 180 °C under the same synthesis condition as those for PbTe NWs as shown in Fig. 1(a)-(c) except for replacing Pb(ac)$_2$·3H$_2$O with Pt(acac)$_2$.

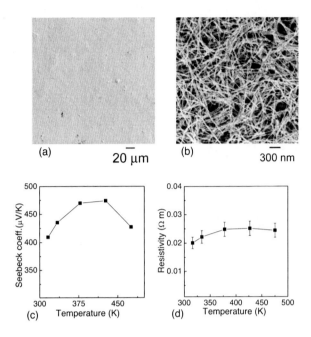

Figure 3. (a) A SEM image of the sprayed PbTe NWs on glass substrates after hydrazine treatment of the NWs. (b) The TEM image shows the morphology of hydrazine treated PbTe NWs after annealing in vacuum at 300 °C for 30 minutes. (c) Seebeck coefficients measured on the film sample of PbTe NWs in the temperature range of 315~525 K. (d) The corresponding resistivity of film sample at T=315~475 K.

Author Index